Molecular Genetics in Fisheries

Molecular Genetics in Fisheries

Bruno Augusto Amato Borges

Delve
PUBLISHING

www.delvepublishing.com

Molecular Genetics in Fisheries

Bruno Augusto Amato Borges

Delve Publishing
224 Shoreacres Road
Burlington, ON L7L 2H2
Canada
www.delvepublishing.com
Email: orders@arclereducation.com

ISBN: 978-1-77469-036-9 (Hardcover)

ABOUT THE AUTHOR

Bruno Augusto is Aquaculture Engineer and Master on Aquaculture and Fishery Recources from Federal University of Santa Catarina (UFSC). His focus is on writing projects and development of new technologies in aquaculture. He is currently project service provider, consultant and develops research on Biofloc Technology (BFT) cultivation for fish and shrimp.

TABLE OF CONTENTS

LIST OF FIGURES

PREFACE

Since the emergence of starch gel electrophoresis around four decades ago, genetic polymorphism has gradually intensified in fish studies. Basically, this polymorphism has been utilized as a marker to make broader inferences about the genetic quality of the research, rather than just including the polymorphic locus itself. Although the introductory research included proteins such as hemoglobin and transferrin, it immediately changed to using enzymatic proteins (isoenzymes). Over the past decade, people have been expanding their attention to coordinate DNA assessment, first with mitochondrial DNA, and then with the establishment of molecular methods, namely nuclear DNA.

This volume suggests that mechanical advances in new genetic technology usually come from natural aquatic system inspections other than fish, despite the fact that this usually requires specialized adaptations to support fish stocks. Fish geneticists need updated information about genetic improvements in model organisms. Sometimes, this method may be misleading but research in fish (such as salmon) can be used in genetic sequencing detection for more reliable outcomes.

To better understand molecular genetics in fisheries, this volume delves deeper into these topics: chapter 1 - diversity and biology of fishes, chapter 2 - molecular genetics, chapter 3 - fish genetics, chapter 4 - genetics and fish breeding, chapter 5 - molecular markers in fish, chapter 6 - catfish hybrid technology, chapter 7 - transgenic fish research, chapter 8 - securing the future of seafood through genetics and genomic application, chapter 9 - conservation biology of fisheries science and chapter 10 - the future of fisheries science.

Since the original study around 20 years ago, the use of genetic methods to solve the problems of ecology and evolutionary research in fish has expanded incredibly. Significant advances in genetic hypotheses, and the availability of genetic markers, cost adequacy, and goals that are critical to the particularly rapid progress in the last decade. The genetic investigation of elasmobranchs has direct significance and has been applied to fishery management and conservation matters, such as defining fish species from their fins.

In the future, the use of modern genetic mapping technologies will grow, which shall accelerate the development of genetic data and the improvement of new markers at a lower cost, preparing for the transition from quality inspections to genome-scale inspections. This volume examines different areas of the submolecular genetic quality of the elasmobranch branch related to fisheries and protection, with particular emphasis on (species distinction evidence, systematic geography, heritability population size, submolecular development rate and reproduction strategy).

Also, the application of genome assets (computational genome sequencing), depending on the rise in the number of sub-molecular markers and the development of high-density genetic maps, sequencing methods can be applied to provide equipment for the genetic development of fish species. In this volume, the quality of various genetic markers and their applications in various aquaculture topics are discussed. Mendelian's genetic predisposition derives genetic standards by noting individual aggregates that appear due to proper breeding analysis. The coordination of data and strategies obtained through advances in molecular science promotes another aspect of genetic quality, that is, sub-molecular genetic quality, which focuses on exploring all parts of genetic quality, such as its structure and capabilities.

CHAPTER 1

DIVERSITY AND BIOLOGY OF FISHES

CONTENTS

1.1 INTRODUCTION

Scientists who are interested in fish today have a fantastic opportunity to research a wide range of live fish, from hagfish to lungfish. They will have a wonderful opportunity to evaluate fish that are suitable for all marine climates (and even some that spend most of their lives out of water). We will deliberately determine how humans live to understand how common animals work and where the fish will exceled to meet human needs. Second, we need to look at these fascinating ways in which fish eyes are formed and learn about the various benefits of in-vitro and in-vivo fertilization methods in different habitats. To highlight the commitment that fish must make, these analogies compare the benefits that fish obtain from the taste of crude oil spills to the benefits that fish receive from human pilfering of their body fluid. In the first place, we should have this, at least so that we can understand clearly what kinds of fish live, how they respond and how they evolve. Amphibian is a species grouping frogs and salamanders with other classes such as reptiles (composed of snakes), birds and mammals. The Mammalia class comprises humans and other animals. With the development of research and scientific knowledge, it is almost impossible to classify new species and genera. Increased extinction will certainly limit them (often our fault by habitat reduction). The class of fish varies from both the class of frogs and the class of crayfish in that... the majority of vertebrates are categorized in a family or group. As of now, some 25,000 different species of fish have been reported. If this number does not occur until the end of the year, it is expected to reach 30, 000 species. Among the teleost species, the good ones overwhelm the poor ones by a few hundred. Cartilage elasmobranchs are rarer, with just a few hundred to a few thousand counted (Fletcher &Rise, 2012).

There are no other species that are as significant, but they may also be of importance in other respects, such as their taste and economic and gastronomic advantages. Within species, all animals share the same general body form. In the Bird class, we see that ostriches and hummingbirds are equally significant. However, both have evolved since the time of the class. They're similar and they're fundamental. In terms of anatomical difference, reptiles can be classified as possessing distinctive features typical of birds, but within the separate, and the distinction between birds and reptiles is merely an option rather than a rigid taxonomic logic. Because both sharks and lamprey are seen, the spectator assumes that they do not seem to be the same in a superficial fashion. The classification of a marine fish as a genus is a reasonably true label indicating a group of fish-like organisms whose ancestry includes an unwieldy relation to the bottom of the sea. On the other hand, to further state Pteraspidomorphi, it is problematic given that each term thus identifies less than a particular group, but represents a morphological stage in the transformation of a group into a not-or, increasingly, a fully aquatic one, 'Marine'. As in all forms of life, most species share common characteristics that mean that they have a mutual ancestor. The hindbrain is a legacy that binds and differs from the common ancestors. In lampreys, all aquatic lampreys, different forms of fish, some newt and some other vertebrate, large Mauthner neurons are located in association with nerve VIII. The hippocampus sends out its massive axon fibers that bridge the two sides of the spinal cord before sending a signal to the various myelinated motor neurons. The lateral hyperstriatum nucleus is absent in the brain cells of all sharks so that it does not grow to all sharks or skip maturity. These two groups of animals do not have a common ancestor at all, and their DNA shows that they have a tighter relationship than gnathostomes do with other groups. The innervation sequence of myotomic muscles of the body warns to impending overload. It is considered a "new" occurrence among all living species but was observed only for the first time, except for one Lizard Fish from the Myotomic group where muscle fibers steadily contract, a phenomenon seen with Lizards other than teleosts who have "invented" a new sequence. Another sign of the nervous system is the contraction of one muscle fiber at a time. This behavior is present in all fish (except sharks and rays) and amphibians.

It is possible to be positive that all the living members of the Pisces class will have a shared ancestor, followed by a molecular flood and morphological hints of this nature. Who was that ancestor, and who was that? -The latest one. Serial muscle blocks or myotomes in the body can be seen to have visible remains in the axial notochord as a chord. It needs to be proved to be true. Some researchers have experimented with very tiny and very delicate live specimens of amphioxus, showing that while it can survive under very low oxygen levels, the genus will also be able to live in non-toxic environments. Paleontologists claim they may have found these unusual fossils, but some of them are still here. Little Pikaia is a small, amphioxus-like fossil fish that emerged in the middle of Cambrian in the Rockies. It has V-shaped myotomes, notochord, and develops a slight bifid head. Pikaia has remained alone for almost a decade (and now seems to be on the front line), but recent discoveries have detected some tiny fish-like chords from impressive early Cambrian fossil lägerstetten fields in the Chinese area *(Cathaymyrus* and *Myllokunmingia*) and are likely to occur. The amphibians are very complex, with a lot of similarities to the fish, but do have a lot of variants of the components found in them (like amphioxus). 540 mia, a group of Cretaceous dinosaurs that carried the first chord-like larynx in shallow waters, breathed and evolved in Cretaceous times when their descendants were around 530 mia. In the simplest form, it is not surprising that all these muscles had certain common features such as notochord chords, which appeared to serve the same function by making the body pass through the same motions when swimming (Bert, 2007).

This means that the muscles most definitely do not belong to the deepest animal class called chords, but must belong to a different group called myochonids. The last molecular attempts to understand amphioxides were informally located in 2006 owing to Holland's applications. There is no such knowledge of any of the oldest chords in the biochemical record dating back to pre-Cambrian times, known as conodons. Calcium phosphate conodontal bones are also used as small fossil tooth-like materials that are known and commonly available to support the purpose of stratigraphic samples. Both conodonts were cut into pieces and examined loosely, but the earlier or later conodonts are certainly a different kind of conodont. Viruses that have to do with these algae or worms can also be present. There is no help for an affinity chord. However, teeth that have a tip that is not yet passed to another culture, which is often induced by the prepared portion of the skin, are also of tremendous geological importance.

In 1983, after the first mistakenly drawn image of the blood vessel was corrected, the images of the internal blood vessels better articulated what the vessels were. Incredibly, the early chordate species were flattened in shape. They were very fish-like with a notochord-like structure, v-shaped myotoms, large caudal fins with teeth, and a large anterior pupil. Any Carboniferous conodont had a distinctive and complicated trilobed arrangement with 3-dimensional skeletons in the gullet of the plant, rather than a mixed set of two-dimensional s-strings with photosynthetic stains. The process that makes up this conodont is definitely nothing like it in any living organism, and we can easily guess-just its works. The link between fossil conodonts and living forms on Earth has not yet been established (some texts suggest a relationship with hagfish, on rather tenuous grounds). An amazing snapshot of life forms fossils from the Cambrian has been given to us. Before the appearance of more modern fish, there used to be a variety of species of fish-like beings swimming with unharmed, sense organs and serial myotoms. All the fish come from one of the chord-groups on the page. This does not mean that all of the fossil fish species that have been found are made up of a monophyletic group, which suggests that they are all inherited from a single common ancestor. In the Cambrian period of 600 million years ago, for the first time, as part of the Cambrian explosion, the first traces of the fish were found in the Ordovician era of 500 million years. Systemists who knew fossil agnathane fish would be related to the form of fossils such as lamproid and hagfish, thereby placing the roots of modern animals in the domain of fossil agnathane fish (Bert, 2007).

Figure 1.1: The Diversity of Fishes: Biology Evolution and Ecology

Source: https://www.google.com/url?sa=i&url=http%3A%2F%2Fwww.brainkart.com%2Fsubject%2FThe-Diversity-of-Fishes--Biology-Evolution-and-Ecology-_260%2F&psig=AOvVaw2-0DBLJBQfj3VCJvP3ynj6&ust=1609427804622000&source=images&cd=vfe&ved=0CAIQjRxqFwoTCKCKj_3_9e0CFQAAAAAdAAAAABAJ

Those closely acquainted with aquatic chords, those fish who are familiar with aquatic chords, experience the same feelings and the same desires, such as thirst, as any land-based person. Although many of the species above the freshwater environment are more common elsewhere, they are more numerous in the intermediate freshwater environment.

The trends of fish distributions are the product of physical barriers that have disrupted the dispersion of past fish and shifts in the temperature shift encountered by classes since the sediment has appeared. You should select the distinct freshwater fish fauna of each continent. When it comes to wildlife, there is a great abundance of species all over the world, but there are scarcely any in polar climates. In the colder regions of the world, the normal species of fish appear to be less abundant (presumably because they adapt to colder temperatures), whereas, in areas much warmer than Norway, these species are more common. There are many different species of fish in the Equatorial Region, but it is difficult to combine them into a certain category. There are many families in the northern temperate and cold freshwaters, including the families of salmonids, the families of sturgeons, the families of smelts, the families of northern lampreys, and other large families, including the families of pikes, leuciscine cyprinids, among others. There is a small range of salt-tolerant species. They dwell in coastal environments, mostly in southern temperate/cold zones. Many oceanic islands, mostly made up of freshwater, have populations of marine species taken from their other habitats (Beamish& Rothschild, 2009).

Freshwater trout, more or less restricted to irrigation channels, is a very cautious method of determining patterns in distribution, illustrating historic continental and climate change dynamics.

1.2 FISH CLASSIFICATION

Except for the primary fish-keeping classes, the classical concept of the relationship between the remaining classes was almost entirely the product of class ontology. At this point of writing, several ichthyologists agree that hagfish and not lamprey do belong to the lamprey stem community. However, there is also a slight number that is not in agreement and suggests the recent location of the lamprey stem group. The names of different species of fish or sea organisms which sound obscure, such as Coelacanths, so the teacher must ensure that students are confident about the key characters of each type. You'd look through these phrases and similar terminology and find bits of knowledge about the fish species – you remember because you've heard about them before (and you know the name) and this extra information goes a long way to make this material more memorable. Any one of the most neglected and nearly obscure was intentionally excluded so that the reader was inspired to do such a search. Also, it is not a smart idea to study the names of the animals if you can do so without discrimination. David Starr Jordan, who gave up the practice of learning animal names, is a very interesting example, and his excuse is that he doesn't want to try to remember the names of unique species he can't know (Ray, 2018).

As evidence to support the argument, some iconoclastic molecular knowledge recently questioned the association between larger groups or clades, indicating a significantly different composition. Scientists conducting biochemical experiments have found that elasmobranches are not basal gnathostoms but are very present in caulifish. Concerning the division of fish species with this predicted upheaval, what can we do? Some chemicals are part of the chemical way of life that could not agree with the current classification of fish but also agree with this classification of sharks. The better way to look at these points of view would be to make more molecular front papers and to give a close review of the data sets to see if and how they corroborate what has been said. Modern anatomical evidence supports that Cephalorhynchiformes and Elasmobranchs have the same fundamental role as normal anatomy. They also had secondary cartilage, which was present in many elasmobranchs. It is not as surprising, however, that these species have risen to the same place as they have historically been. Creator Erik Stensiö justified the presumption that our ancestors were exclusively vertebrates, regardless of the most recent molecular evidence of the fish, could still be true. In a well-written article on this topic, Miles presented strong author comparison statistics and had a fair debate throughout that article. The trouble with fossil fish fossils is that the room does not permit the identification and clarification of their relationship or fossils (a new revision by Forey should appear before too long). Based on the chord-to-chord body size comparison, we can see that protochords have greatly increased genome sizes, especially in the Chordata. The fact that the first jawed fish (and later gnathostomous) was octoploid, but the next gene, regardless of the mechanism, has increased physiological complexity could help to explain the evolution of such complex physiology. This is a mystery; how this could have happened (Ray, 2018).

Cladistics

While classical classifications have usually agreed on the relationship between the remaining classes over the years, there is more argument between the different classifications of the relationship between these classes. Most scientists have said that the hagfish stem is not a genuine lamprey and is best known as a relative lamprey. While a little research has been undertaken, there are not those who agree that stem cells have a recent standing. Names of unique classes of sea organisms, such as the Coelacanths, which are elusive enough to ensure that the teacher has faith in the main features of each type. You will work through these sentences and find if it is the name of the fish species that are being tossed that provides an emotional atmosphere that makes the sentence memorable, and it is those related sentences (i.e. statistical statistics on the fish species) that provide you with the details that will make the fish species unforgettable. All the subjects selected by the filtering process have been skipped, allowing the chosen themes a hunt of this kind. Essentially, it is not prudent to learn the names of the animals only if you can't do anything without prejudice. David Starr Jordan, an American physician, had stopped studying animal names for a time because he could not remember enough of them. In his defense, he said that those names would not be of any concern to other citizens.

Recently, it was suggested that such compositional units may be distinct from others, and in support of this, iconoclastic molecular information recently contested the association of the units. Almost all of the biological findings surrounding elasmobranchs suggest that they are in gnathostoms rather than invertebrate communities. In this article, what does the author mean by this "expected upheaval" about the "disparity of population" Stuff like toxins are often classified as part of the biological climate, and may or may not adhere to the mark of fish but often agree with the classification of sharks? The best way to look at these points of view would be to make more molecular front papers to take a closer look at the data sets to see whether and how they help what has been said. The basic characteristics of the organs of the present anatomical proof of this gene are nearly identical to those of modern humans. Secondary cartilage was found on several elasmobranches. Of course, we are not surprised that this modern mammal species has returned to its former role. To legitimize his argument about the role of vertebrates in the modern world, designer Erik Stensiö has defended the quality of our ancestors based on up-to-date genetic data from fish. In a well-written article explaining the author's observations, Miles offered an intelligent description of the topic and had a very well-thought-out discussion. Many fossils are not classified into some kind of family, since they do not have enough data to classify them. When we compare chords, the two main differences between both of these chords are the greater chromosome sizes and the fact that they are vertebrates. Evolution has likely conditioned us to understand what is happening by understanding the first vertebrate, but evolution has made physiological developments in the next gene independent of the process. This is the most wonderful of all the puzzles.

1.3 GLOBAL SPECIES AND TAXONOMIC DIVERSITY

1.3.1. Fresh, brackish and saltwater fishes

Ichthyologists have been able to distinguish three major classes of freshwater fish due to their aversion to saltwater and their presumption that they prefer to migrate through coastal landscapes: principal, saltwater-intolerant fish; secondary, often capable of breaching small marine barriers; and peripheral divisions, mostly made up of aquatic fa par. Trying to catch three different types of fresh and brackish marine life with very detailed fish information makes it difficult to keep up with the sources.

- Unique freshwater.
- It has been found in fresh and saltwater.

It may exist in natural, tidal and brackish waters. The primary division is one that is more or less the "main" division, and the second division is the "secondary" division in grades 2 and 3.

Figure 1.2: What's out there? Diversity in the Ocean

Source: https://www.google.com/url?sa=i&url=http%3A%2F%2Fblog.cwf-fcf.org%2Findex.php%2Fen%2Fwhats-out-there-diversity-in-the-ocean%2F&psig=AOvVaw2-0DBLJBQfj3VCJvP3ynj6&ust=1609427804622000&source=images&cd=vfe&ved=0CAIQjRxqFwoTCKCKj_3_9e0CFQAAAAAdAAAAABAj

1.3.2 Global estimates of fish species diversity

At this time, we are still not sure of the number of species of fish that exist. Although more and more information has been known about 478 species of Teleost fish since the early days of the 19th century, our general understanding has vastly improved.

After the inventories, the fish catalog established and predicted some 27,300 true fish species, with an estimated 31,500 species. The number of species on the list had increased to 21,000 in September 2005. Of the 28,000 freshwater and marine species in 2006, 48 percent of the 54,711 recognized living vertebrate species were listed in the Illustrated Encyclopedia Book.

Under the above scheme of different species evaluation, the classification of Eschmeyer is followed by the Fish Base regardless of the degree of difference with the Nelson figures. Although the number of fish species is increasing in freshwater and brackish water environments, the uncertainty of the scales of 'brackish species' can help to clear up the increase in fish species that use freshwater Habibi (15,062 according to Fish base).

If 33 percent of fish screened less than 5,000 ppm were not present in the saltier regions of mangrove estuaries, between 40 and 43 per ent of all fish would be found in freshwater. The supposed awareness of the most recent findings of fish accounting for an even higher figure of 45%. Notably, whatever the exact number, only 13,000 freshwater animals exist in lakes and rivers, accounting for just 1% of the planet's biodiversity, while the remaining 16,000 live in marine ecosystems, accounting for 70% of the world's land.

1.4 TAXONOMIC DIVERSITY

There are 207 species of pelagic fish in fresh and brackish aquariums (170 for freshwater fish in particular). Species C (characiform, cypriniform and cyprinodontiform, Ostariophysans, Ostariophysans, Cyprinodones, and Myttiloidei) and species D (Cyprinodons, Siluriformes, Gymnarchidae, Gymnotiformes, Siluriformes, Hemiramphidae, Hemiramphidae, Gymnotiformes, Hemiramphidae and Cyprinodontiformes ostariophysan) may be readily described. The variety of taxonomic families shown below is subjective, but still gives a strong example of the taxonomic equity of families (i.e. species). The Neotropical region consists of two large regions, the Amazon River Basin and the Orinoco River Basin, both of which are crossed by the Colombian and Orinoco Rivers, forming two major watersheds. These watersheds cover 93.1% of the landmass of the Neotropical zone. There is little surface water in these large basins, and much of the water met is turbid (i.e. primary and secondary divisions).

Freshwater and brackish fish numbers are available for 4,231 species(769 genera) in the Neotropical Region, for the Afro-Tropical Region, for 3,272 species (401 genera) in the Eastern Region, for the Palearctic Region, for the PA Region, for 2,381 species (551 genera), for 1,741 species (402 genera) in the Nearctic Region, and the Palaearctic Region, for 5,801 (55) genera (1,232 genera).

Australian

The global fauna of fish has some general characteristics.

If you are interested in seeking reasons for the enrichment of habitats in various regions, a variety of different hypotheses have been suggested. The first "area" theory suggests that the abundance of biodiversity as a feature of its surface region is increased by the size-dependent rate of extinction/colonization and/or enrichment of the ecosystem. Secondly, it is important to note that the productivity principle demonstrates that the abundance of the biodiversity of the region is positively related to the total resources available. Third, the patchy pattern of variation uncovered by fossil records is not replaced by a standard distribution of traits but is often replaced by a patchy pattern of extinction due to historical contingency in which species undergo speciations at rates that are unevenly distributed across the globe. As far as freshwater fish are concerned, it is clear that the area itself clarifies a significant proportion of the variability of capital. Species diversity increases over time in rivers and habitats, and there is not much to discern between regions (basin area and volume of discharge). However, energy supply and history often explain a large part of the trends of income found, taking into account the industry.

Figure 1.3: The Freshwater Fish

Source: https://www.google.com/url?sa=i&url=https%3A%2F%2Fwww.technologynetworks.com%2Fapplied-sciences%2Farticles%2Fthe-freshwater-fish-crisis-316375&psig=AOvVaw3XCA3QRD-KfMMbLLFfl9G-&ust=1609448519353000&source=images&cd=vfe&ved=0CAIQjRxqFwoTCMCb8I3N9u0CFQAAAAAdAAAAABAb

Phylogeny and historical processes

In the water of the world, the current distribution of freshwater fish has undermined millions of years of global changes in the water cycle over the last decade. No other aspect impacts the stability of freshwater environments, including temperature changes, and the freshwater surface environment has been altered and dynamically modified. As a result, many marine ecosystems have been transient, their ecology has shifted, and their aquatic organisms that have survived in legacy systems of differing species ratios and their developmental, middle and later life stages, reproductive systems, migration mechanisms, life strategies and adaptive responses are likely to survive again under current conditions. It's about looking at the atmosphere and volcanic activity and trying to see when they happen and the issue of how a newly introduced animal species could spread the disease and build a new population revolution.

More than 350 million years (Myrs) , fish are found in the center of the Devonian continent. These fish have been categorized into a few different groups: jawed fish (Chondrichthyes), , Dipnoi, and some chondrostei. Many of the individuals who had this viewpoint firmly disagreed with the pro-continental drift perspective.

Gondwana, its less tropical regions, and Tethys' descendants are now living in freshwater. Indicating the presence of Gondwana as a whole to the members of the lineage. Many eagles from around the world have been found to have existed for hundreds of thousands of years. This is the case of "Dipnoi"

A single-pin group that was first formed in Devonians and diversified in the Mesozoic, and is currently distinguished by a few well-known species in Africa, Australia and South America, as well as a distinctive teleostic cover of the late Jurassic stage Osteoglossomorph, which divides its subgroups across tropical freshwater regions with North America.

The Polypteridae family is diverse and very common. They are numerous, but only present today in Africa and the fossil record. Polypterides are also widely spread.

This group is very rare and monophyletic. The history of the biogeography of the Otophysus in the 20th century has been under consideration, particularly since its first fossil record in the 19th century. Asia has long been regarded as the place of origin for the Cypriniform, where the population is the most specious. The data now reveals that in the late Jurassic (approximately 150-160 Myrs) Portonarthas origins in South America and the Tethys Sea, and throughout Asia in the Late Jurassic (approximately 150-160 Myrs) and evidence of cypriforms and siluriforms emerged in South America. Even before Laurasian mammals split into Laurasiatheria (South Africa) and Gondwanothere (India/Pakistan), Diogo suggested that South America was the place where catfish origins could be formed. Some catfish soon encountered new worlds and traveled from the East to other continents, while others traveled to Laurasian regions.

In recent years, paleontology analysis has shown that many marine animals have disappeared from South America during the Cretaceous era and since the Cenozoic period. Although the Miocene era was full of diversification of species, the Cretaceous period was where these fish were very diverse and diverse. Late Miocene fish fauna was populated by new species (10 Myrs). The distribution and variety of the species of fish continued even after the conclusion of the taxonomy. As historians have recently learned, the overwhelming amount of gender and family diversity over millions of years has not been triggered by volcanic activity.

Continental distribution: the major features of Ithyofauna in the continental ranges.

North America

The Transvolcanic axis just south of the Mexican Plateau is situated in North America from Alaska to Canada, taking around 1,050 aquatic freshwater animals around the globe. The North American continent alone ranks seventh, with more than 800 distinct species on the mainland, and is considerably richer in species than other temperate areas. This type of fish is not yet widely spread like it is in other parts of the world. Historical factors, such as the impression of previous glacial cycles, have had a long-term effect on the history of the watershed, local geological and climatic influences on the landscape, hydrology and biogeography. There are more than 500 species in the south-eastern United States, but the priority may be on endangered species in the southern Appalachian uplands, the country's richest species region (except maybe the Florida Keys). From the east to the west of the south-east, we can see that continental biodiversity begins to decrease significantly. The combination of influences, including a diversity of physical geography, a favorable tropical climate and a long but distinct history of zoogeographic activity, is undeniably the product of the exceptional richness of Southeast America. While the remainder of the continent saw successive freezing and retreats during the Pleistocene, this area remained relatively untouched, enabling the environment to steadily evolve into today's animals. The tropical southeastern part of the U.S. is made up of a wide diversity of plant and animal life, while the mid-eastern and eastern parts are made up of a greater variety of exotic plants. More than 250 species of plants and animals are known to live on the Mexican plateau. In these, some 200 belong to a single-family, one of which is a diurnal family.

Main characteristics

35 Northwestern and Nothern Waters families remain in North America. 13 of these families, all of them native to the area, live in NW and NE Waters. In all fish families, the composition of Cyprinidae, Percidae, Poeciliidae and Catostomidae (which make up the majority of fish species) is more than 50% of all fish species in general.

The drainage of North freshwater creates the flow of freshwater that flows into the water to create river systems periodically (e.g. Cottidae, Atherinidae, Clupeidae, and Moronidae).

American freshwater fauna

Anadromous trout and salmon are dominant in the freshwater ichthyofauna populations of the Pacific coast during the spawning season. They mediate a major change in nitrogen from the ocean to freshwater regions. Since Brevoortia and Alosa are distantly related, they have at least a great family history of producing populations that were part of the Atlantic coast and inland seas.

This species is found in the United States and much of the world's lakes and rivers, as well as in the few remaining habitats, such as the Great Lakes of the United States, where it lives alongside alive, enigmatic bowfins.

In North America, fish species are mainly Eurasian-like elements such as Esocidae, Umbridge, Cottidae, Cyprinidae, Percidae and Gasteosteidae, or European-like elements such as Cyprinus, but Neotropical elements are also clearly described, especially in southern regions such as Catostomidae or Rivulidae.

In 2000, Dr. Karel Smith performed a very accurate and strong analysis in Lundberg on the diversity of fish and northern tropical regions relative to the well-recognized tropical regions and species diversity. A few new species are found and identified annually, but the scope for substantial biodiscovery is likely to be confined to southern Mexican regions and North American Ithyofauna in general. Mayden et al. (1997), in addition to numerous regional works on Ithyofauna and the comprehensive treaties of 1997 have achieved excellent performance.

South and Central America

The Neotropical field involves the species fields of South and Central America. The world with the most diverse fish population can be found in freshwater (around 3,600 freshwater fish species). Owing to the similarities between the Neotropical Fish Fauna and the Nearctic Fish Fauna, the Afrotropical region is not particularly close to what is found in North America.

The fauna of the Central American region is a mix of North American and South American lines from the Tehuantepec isthmus to the frontier between Panama and Colombia. There are native cyprinodontic organisms and there are other secondary peripheral members of the tribe. In both Central and North American freshwater fauna, only two North American families of frogs have been recorded (Catostomidae and Ictularidae).

Main characteristics

The vast majority of Neotropics consists of five broad groups of fish: the Characiforms (about 1500 species described, probably 2,000 species), the Siluriforms (1400 of the known species, possibly 2,000 of them, the Gymnotiforms (about 180 of them), the Cyprinodontiforms (about 400 of them) and the Cichlids (about 200 of them) (some 450 species). Seafaring characids and siluroids are so unbelievable that people didn't even know they could survive there.

Succumbing to the decimation of oceanic species by industrialization, South America is utterly destroying all of Africa's primitive fish families, save for the Lepidosireniformes and Osteoglossiformes.

Local cyprinids can't be found in this lake.

The Gymnotiforms are African and Asian fishes that possess electroreceptors such as rays but lack a swim bladder and are very unusual as well as swim bladders.

The proliferation of Neotropical fish, small species including characiform, siluriform and cyprinodontiform, tends to be representative of the Neotropical fauna of fish. I've heard of very large fish like the 90-pound Arapaima gigas (up to 3 meters long and 140 kg) and the 70-pound

Brachyplatystoma goliath, but they're just pups compared to those very large fish in a big group (up to 4.5 m long and 200 kg).

Over 800 new South American freshwater species have been described over the past two decades and the pace of classification is projected to increase, based on living conditions and the size of the lakes. The prediction on the collection of some new Neotropical species was projected to be roughly 8000. Lundberg estimates for Neotropical invertebrates of about 5,000–8,000 species. Today there are only around 4500 animals, but several more have been discovered recently. Several species are still waiting to be called. There is no systematic study of many fish fauna in the Neotropical region at present or over time. A lot of literature on the subject was identified at the symposium, which was very useful to the research.

Europe

Europe's fish fauna was in a much worse state than other continents during the dry season. The fauna of freshwater fish. While the glacial areas (distances) were re-colonized during interglacial and post-glacial times, they emerged primarily from the Ponto-Caspian region and, in particular, from the central and lower parts of the Danube basin, which acted as a large shelter from most climate change. During the last ice age, some of the earlier Europeans also moved to other regions of Europe as a result of climate change. There are also signs of this "re-colonization" where we might see some of the northern fish as isolated during the final phase of the last ice age (the Pyrenees, Alps, for example). Elevation, latitude and temperature are also variables responsible for endemism in the southern regions.

Main characteristics

The largest group of giant freshwater fish is Cyprinidae, which is made up of about 20 families, but about half of the species are Salmonidae, Coregonidae, Gobiidae and Cobitidae.

There are few endangered species in the areas that have acted as the primary refuges in the last ice age, and the greatest numbers of them are present.

The Danube River is one of the major rivers of Europe and has the most abundant fauna on the continent.

By the beginning of the 2000s, the number of native freshwater fish species in Europe had grown to 250, and by the end of the 1990s, to around 360. The Russian River in the western part of the former Soviet Union had more than 400 native freshwater fish species. The taxonomy of this and the other salmonoid families is of considerable benefit in identifying the specific species that you are seeking to catch.

Africa and Madagascar

Billions of years ago, around 600 Myrs ago, sea levels on the African Continent were even higher. This includes the regions of Niger, Sudan and Somalia, though parts of Ethiopia have also been found in the ocean. As a result of the discovery that we have had for such a long time, we have many African fishes today that can grow old and live incognito. The whole of Africa is such an amazing place.

Main characteristics

There are about 48 freshwater fish families in tropical and southern Africa. In total, 15 of these families are endemic. While the number of family groups in Africa has declined, Basal and Archaic families in Africa have been able to thrive as a result of climate change.

Most of the groups of Chondrichthyes includes polypterides (sister group of Chlamydiale), phractids (sister group of Cladobatoidea), knerides (sister group of Carcharhiniformes), chromerides (a group of Cladobatoidea) and graseichthydids (sister group of the Temnodontiformes).

African fauna also contains traces of extinct elements of larger distribution, such as Proteropterida, Notopterida and Osteoglosida. These species are contained in the sediment of Africa. Other than Pantodontidae monotypes, Mormyridae and Gymnarchidae and Pantodontidae monotypes are the three other endemic families in Africa.

Two main freshwater fish lineages, the Cyprinodontiformes and the Cichlidae, are present in Africa. Quite diverse, these two lines of fish have very different characteristics.

For most inland-African families, the fish that make up the tribe is comparatively smaller than other families. The Clupeidae genus is also known as the Ostracods. The Ariidae family is also known as "Mudpuppies" Family Synbranchidae is sometimes referred to as "Tubenoses" The Centropomidae family is also known as "Aardvark" The Latidae family is also known as "Hippos" The Gobiidae family is also known as "Molehippos" The Eleotridae family is also known as "Elephants." The Mugilidae family is also known as "Sloth" The Syngnathidae family is also known as "Puffadder" The Tetraodontidae family is also known as "Tetraodons".

Diadromal rocks are rare in the African continent. The Maghreb and the eastern coast of the continent of Africa are home to five distinct members of the Anguilla genus. Yet, in West and Central Africa, the genus is missing. We see this kind of thing in other nations (e.g. Sicydium).

Madagascar is similar to the freshwater fish fauna on the continent of Africa. Of the 135 freshwater fish species, only 84 are not native to this island. A few new bird species have recently been identified and, as I am well aware, many more are yet to be discovered. The fauna of freshwater fish is still unidentified. The vast majority of plants are found in the suburbs and are typically scattered. As one journey around the African continent, it can find that not all freshwater fish species live in the aquatic areas of other regions of Africa (anabantoids, bagrids, clariids, mastacembelids, noopterides, etc.).

There is a long waiting list in some rivers for the major freshwater families, such as cyprinids, characters and mormyrids. Three of the four families on the island are allowed to grow calves (Aplocheilidae, Cichlidae and Poeciliidae). The Malagasy species members of the phylogenetic family are, by consensus, ancestral to their ancestors and the invertebrate is distinctly monophyletic.

Three thousand five hundred and thirty-five species of African freshwater and brackish fish are listed at this time. There may be a variety of different meanings that apply.

Asia

This region covers land east of the Indus River and south of the Yangtze River Basin. The region contains many large islands, such as Borneo, Sumatra and Java, as well as many smaller islands and the mainlands.

Main characteristics

High numbers of classes of fish: 121 from inland waters, including 34 primary and (or secondary) freshwater areas (18 from Southeast Tropical Asia) and 87 from outlying waters, typically just a few species. The most popular are Cypriniformes, such as Cyprinus (approximately 1 000 spp.), Cobitidae (approximately 100 spp.) and Balitoridae (approximately 300 spp) (85 spp.).

The central family lives in the oceanic basins and islands, while the peripheral family lives in the oceanic river basins and islands. The rest of the region of Sulawesi and all of the Moluccas. While some of the islands in the Philippines are mountainous, there are no traditional freshwater habitats. The Gobiidae is the main peripheral group (about 300 spp.).

Besides being miniature organisms, many of these species mimic South American animals. Borars in northern Thailand are adults with a weight of around 12 mm SL and a length of 7.9 mm SL. It is a sexually mature Pedocypris progenetic. The greater one of these species dubbed the "giant stingray" has a depth of 2 meters, on the other hand, a height of 6 meters and a weight of up to 600 pounds. It is said that Pangasianodon (Pangasius) is the largest freshwater fish in the Mekong River with a recorded weight of up to 300 and 3 million pounds.

Currently, we don't know how many species there are, but we do know that the number is estimated to be larger than previously believed. The Fish Base reports 3,553 species of freshwater fish, or 4,400 species, whether we count fish on the edge of freshwater.

Knowledge of tropical Asian fish fauna is still being studied, especially in China and India, where surveys are not complete. Several species need to be identified or recommendations need to be made.

Oceania

To be represented as a "biogeographical entity" Australia, Tasmania, New Guinea and the Australian continental shelf must have well-defined boundaries. Another domain, like New Zealand, is approved.

Approximately 100 Myrs ago, land relations between Australia and other continents (including Southeast Asia) were separated. The myrtle's soil has been in a visible dry state for the last 15 years and has led to the draining of the river, mostly in the palaeocene. Every trend of distribution, within the period of the Miocene, has almost definitely evolved in the distant past. The last large glaciation was very moderate and had little impact on the ecosystem of aquatic species.

Main characteristics

There is an impoverished freshwater fish fauna in Australia, and not all large Otophyse freshwater families are present elsewhere in the world.

Although most freshwater fish are members of marine families with some electric fish groups.

They are the most commonly used freshwater fish present in Australia and New Zealand. Many of these animals are bony fish, although some have become extinct in modern years. The species was present only in Queensland, home to swamps and permanent rivers. There is also a considerable number of endemics in the southern, central and western areas of the continent in separate parts of the continent. There are, however, few in the north-eastern portion of the continent. The division of arid and rainy areas which have much to do with the kind of climate in which this species lives.

In NZ, seven fish families (Galaxidae, Eleotridae, Anguillidae, Cottidae, Poeciliidae, Anablepidae, and Ictaluridae) are not expected to be present (on or after 11/7/15) in the country's waterways. No freshwater fish are significant, and the diversity of low species in New Zealand is not high. About 60 percent of New Zealand fish populations are diadromes. The number of species found here is scarce.

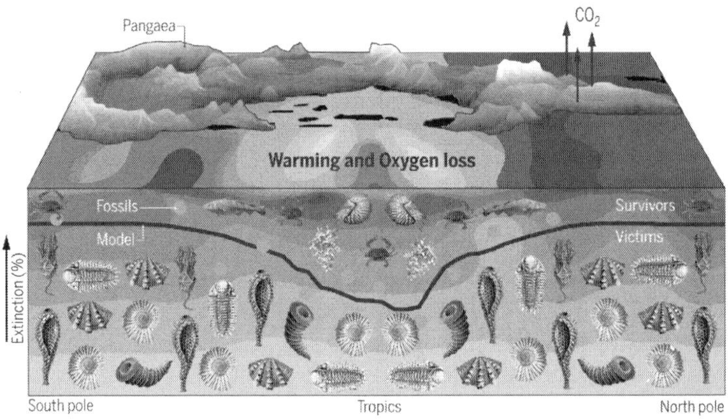

Figure 1.4: Biggest extinction in Earth's history caused by global warming

Source: https://www.google.com/url?sa=i&url=https%3A%2F%2Fwww.washington.edu%2Fnews%2F2018%2F12%2F06%2Fbiggest-extinction-in-earths-history-caused-by-global-warming-leaving-ocean-animals-gasping-for-breath%2F&psig=AOvVaw3XCA3QRD-KfMMbLLFfI9G-&ust=1609448519353000&source=images&cd=vfe&ved=0CAIQjRxqFwoTCMCb8I3N9u0CFQAAAAAdAAAAABAh

Main ecoregions in North America

In 2000, Abell described ten North American Regional Complexes, mostly focused on distribution data for freshwater trout, mouse and crayfish. This comprises the Appalachian Highlands, Laurentian Mixed Woodland, and Great Lakes-St. Lawrence Lowlands, Mountain West, Western Cordillera, Pacific Coastal Steppe, Pacific Dry Sagebrush Steppe, Red

Deserts, Columbia-Snake River Plain, and Northern Pacific Steppe. From the documents we have read, it looks like squid, mouse and crayfish are abundant in North America, and we will remember the Abell base map as a replacement for North America's geographical division in the provinces of Ithyofauna.

- Maritime Pacific Province (ca. 40 endemic fish species). The country comprises mostly of man-made bodies of water, coastlines and rivers. The restricted desert ground is found in the Sonoran and South Pacific areas.
- The Great Basin is situated in the province of (23 endemic fish species). The set of water features also comprises water bodies, lakes, rivers and springs.
- PLACK, WHITE, OR Gray (20 endemic species). A combination of reservoir water, Colorado water source, and more from many reservoirs and creeks.
- Province of the Grande River (ca. 80 endemic species). This is the region of the Rio Xeric and the Upper and Lower Rio Grande, which is created by the rivers, and the mosaics formed by the springs.
- Mississippi State (ca. 130 endemic species). Mississippi is the number one high source of irrigation in North America, the highest in the USA. There are more than 375 species of water-dwelling fish that rely on mainstream tributaries, headwaters, and water bodies to survive. This area was a place where the Reindeer passed, and there was a place where the early humans overcrowded.
- it's the province of this continent (ca. 65 endemic species). Cape Cod Mosaic consists of subtropical coastal canals, wetlands and springs to the south, spreading northward to the colder, smaller rivers and its cooler tip waters and lakes.
- Saint-Laurent Province (at least three endemic species). The comparatively mild Gulf waters of Northern Nova Scotia, the St. Lawrence River and the Grand Lake have permitted a growing number of aquatic species to find a home.
- Hudson Bay Province (no endemic species). Southern Canada's headwaters and streams stretch east of the Hudson and James Bay, ending at the Arctic Rivers and the Hudson and James Bay Lakes.

- Province of the Arctic (four endemic species). There are the Arctic Yukon and the Arctic Alaskan Rivers and the Arctic Archipelago.
- Mexico's Bioregion is now undergoing transition (ca. 200 endemic species). Xeric seasonal canals, lakes and springs, compared the plateau to the rare natural beauty of the desert island.

Main ecoregions in Central America

In Central America, more than 350 species of plants can be found between the Tehuantepec Isthmus to the northeast and the boundary between Colombia to the southwest. Four provinces or lesser different regions have been defined for each subfamily of Ithyofauna. Due to the shared geographic range and population, these provinces have been divided into four main regions.

- The goal of this study is to begin in Usumacinta, in southern Mexico, along the North Atlantic and North Pacific rivers that drain the current country.
- The Pacific coasts of Costa Rica and Mexico form an important part of the Pacific Province.
- The Atlantic River system of Nicaragua and Costa Rica is situated in the province of San Juan.
- The Isthmus Province exchanges various types of trees with the Magdalena Province of South America, Panama consists of the Pacific and Atlantic rivers of Panama and the other Pacific rivers of Costa Rica.

Main ecoregions in South America

For South America, the analysis of distribution and research on the role of fish on the basin scale in the case of 10 provinces in Ithyofauna is considered to be provisional based on a preliminary report.

- The Province of Southern Patagonia (12 species, 1 endemic). From the southern tip of Patagonia, all the way to the Chubut River, the southern rivers of Argentina flow.
- The province of Patagonia in Argentina is Colorado and Patagonia in the provinces of Colorado and Argentina (23 species 5 endemics).
- The transandinian (south) regional area is also responsible for all remote, coastal river systems in Chile (19 species, 13 endemic).

- Lake Titicaca Province (32 species, 30 endemics). The Bolivian and Peruvian Andes Endorheic Drainage appears to be supportive with *Orestia* spp. In the highest and most isolated areas.
- The Paraná Province (847 species, 517 endemics). The major rivers of La Plata drainage and Sao Francisco drainage in Central Argentina are primarily coastal rivers along with islands.
- It's from Brazil. –South America.
- Southern Brazil is made up of several tidal streams, and the Lagoon system is in the south-east of this region (194 species, 90 endemics).
- Brazil's SE Province (131 species, 50 endemic) is a large part of the eastern coast of Brazil. At this point, the data is not abundant.
- The Amazon Province is a region of South America (2,416 species, 2,072 endemics). One of the biggest and most underdeveloped nations in the world is mostly made up of rivers, such as the Amazon and Orinoco, and lake interfaces, such as those on the Shield of Guiana. Several discoveries are anticipated in the future.
- Province in the North of Venezuela (61 species, 9 endemics). Province of Magdalena (north) and Trans Andes (south) (423 species, 326 endemics). The Ecuadorian Rivers, the Columbia Rivers and the Maracaibo Rivers of Venezuela are all/part of the ETS.

Main ecoregions in Europe

In 2005, Reyjol published the most recent data on the European atmosphere in the basin. This study used a list of species from 406 basins (233 species) that were reasonably uniform in abundance in Europe and used it to classify the regional (provincial) regions with a homogenous fish fauna.

- Iran's Islamic Republic (98 species, 36.7 percent endemic),
- Northern Europe Area (42 species, 9.5 percent endemic),
- The native plants of Western Europe are mostly found in Western Europe.
- The Middle Province in Europe (57 species, 1.8 percent endemic).
- The Central Province of Peri-Med controls much of West Africa (93 species, 64.5 percent endemic),

- Improvements are being made in the Eastern Mediterranean (64 species, 31.2 percent endemic),
- The Iberian Peninsula, to use the proper word (50 species, 60 percent endemic).

Main ecoregions in Africa

Several African countries have been set up:

- Fauna in the Maghreb was thought to have unhealthful Paleolithic origins.
- The Sahelian drainage system comprises the main Sahelian drainage basins: the Niles, Chad and Nigen, the Senegalese and the Volta. The region of Sudan from the Atlantic to the Indian Ocean is called the "Nilo Sudan." Sea fish are now very abundant.
- The coastal streams of Guinea, Lower Guinea, Upper Guinea and the Congo Rivers of West Africa include the waters of Guinea, Lower Guinea, Upper Guinea and the Congo. With several native taxa, the fauna in the area is well adapted.
- Coming from Cameroon to the mouth of the Congo, the western coast of West Africa is dominated by the western coast of Central Africa and its large fauna.
- The Congo Basin, located in the east of the Democratic Republic of Congo, occupies the same province as the Congo Basin. It is unbelievable how diverse Ithyofauna is, but it is difficult to manufacture a synthetic material and so many different technologies need to be made.
- The province of Morro Branco can be identified as "extremely poorly defined" by the Angolan Coastal Drainage Area.
- The Zambezi is a tributary of the Congo Nile, and this river has a wide variety of fauna, such as the Cunene, Okavango and Limpopo rivers.
- The Juba-Zambezi tidal wave takes much of the water upstream to the East Coast of the USA. The fauna here is relatively abundant and is searching for a new perspective.
- The southern region of the planet consists of the Orange/Vaal Basins and the whole southern water network. This fauna is a little copious and common.

Main ecoregions in Australia

Unmack has recently established the 3rd largest collection of e-cigarette smoking communities. Most of the habits in question were developed in different periods, possibly as early as the Miocene epoch. Polio-Pleistocene extinction seems to have little or little effect on the large distribution patterns of freshwater fish in the world.

- Based on the physical features of the environment, there are a few species and none are apparent anywhere that suggests a remote region. Fauna relations are farther away and southeast, respectively.
- The Pilbara region is one of the warmest regions of Australia. The species in the "Miniopterus" group are present only in the Southern Highlands and Lowlands of New South Wales.
- It's a very prevalent species in the Kimberley province (of South Africa). Out of the 29 species, only 16 are common. (Reframed)
- There are 38 endangered species among the 755 animals in the Northern Province (38 are endangered) (50 percent endemic). In New Guinea, the rivers the Nile, Fly and Murray are all very similar, and all of them contain 33 out of 75 fish (45 percent).
- Humans and other animals are suspected to be extinct in the eastern province due to the intense lack of food and major changes from the north. Of the 47 species, 15 are endangered (31 percent).
- The Bass Province's Basic Fauna.
- There are 8 endemic species in the southern province of Tasmania, and 3 universes, three of which are paragalaxies, and no regular monsters. Many of them have small fields surrounded by one or two lakes and/or streams.
- Similarly, the Murray-Darling region seems to have acquired its fish and other species from surrounding coastal regions, although they still maintain a degree of endemism.
- In Central Australia (30 distinct plants), there are 10 varieties.
- To the north of Palaeo, the drains used to flood but had dried up since 1993. As of now, it's been only one isolated species from this vast region.

Endemism and fish species flocks in ancient lakes

There is a remarkably high number of fish species in some ancient lakes, some of which are very ancient (overgrown) like a thousand years old. Among the sea birds that come from a single lake to breed, the greatest and most interesting characteristic is the appearance of large flocks of the same species, which are very old and native to the area. The stability of old lakes could be the explanation for the excess of exotic evolutionary radiation that they maintain relative to younger lakes. When it comes to intralacustrine speciation, there is an increasing number of signs that there might be additional speciation beyond intralacustrine allopatric speciation. It has recently come to light that many distinct species of fish today form populations in numerous ancient lakes, which are excellent natural laboratories for speciation analysis.

Africa is marked by its own highly endemic Cichlids, which seem to have evolved differently from the ancestors of the river. There are roughly 550 haplochrome cichlydia natives on Lake Victoria, possibly over 800 haplochrome cichlids natives on Lake Malawi, and 325 haplochrome cichlids natives on Lake Tanganyika. In the Congo, in the Lakes (Kivu, Albert, Edward and Turkana) and the waters of Cameroon's Crater Lake, local cichlids have been sifted to a lesser degree in small fry. This hypothesis suggests that certain cichlid species of 100,000 years of age or older may have a high life span (Shimizu et al., 2013).

The non-cichlid fauna of the eastern African lakes is minor and out of the ordinary, but it is also less common in the area. Of these other families, there are more than a dozen members who are mostly small flocks. For eg, there are seven species of mastacembelin and six species of chrysichthys.

A large population group of labeobarbus species has recently been identified in Lake Tana, Ethiopia. N. Nagelkerke & Sibbing are currently performing detailed morphological research, reproduction and feeding of *Labeobarbus intermedius* complex species in Lake Tana. At least 15 animal species, which are biologically diverse and vary in the niche in which they live, the plants they eat and the breeding land in which they live, have been described. The 24 species of Orestia in South America are being researched as they remain in Lake Titicaca (Cyprinodontidae). Lake Chapala is a small lake at the bottom of the mesotrophic chain of lakes, where 8 silverfish fish predominate. It is also said to have a colony of "silverside"-like fish for privacy purposes. A flock of six species of Cyprinodon (pups) has also been found in the Chichancanab lagoon on the Yucatan Peninsula (Mexico).

The lakes have existed their entire life (at least 60 million years) in the eastern part of Siberia, around the Baikal Rift (a deep volcanic rift within the Earth's surface). The present lake has created one or more large lakes over 28 million years of its lifespan. Today, Lake Baikal has a very rich biodiversity of 2,500 distinct species of mammals, 56 different species of fish, 15 different species of amphibians and 28 different species of flamingos. In particular, the presence of a flock of birds (Cottotoidei), which includes 29 species (11 genera). Life has continued to colonize the most regional ecosystems, including the abyssal and pelagic regions of the lake, across a variety of adaptive radiations. According to the recently completed genetic analysis (bone as characteristics representation), several sculpoid fish inhabit the region, suggesting that the ancestors of these fish are thought to have been created before the end of the Pleistocene period when temperatures in the world were much higher.

Lake Lanao National Park has been developed 3.6 to 5.5 years ago in the tropical Asian region. In this case, it can be argued that animals are becoming capable of adapting to new environments. Unfortunately, this population has been decimated by exploitation and experimental introduction of fish all over the world, and only a few viable individuals exist today. 26 endemic plants create a variety of palatal lakes in the central part of Sulawesi (Indonesia) (22 endemics). Tectonic Lake is the deepest lake in Japan in the middle part of the country. Recently, Old Biwa Lake has moved from a shallow to a moderately deep lake. It also shows that the Reservoir, which was formed a very long time ago, started to shape around 300,000 years ago. There are 71 species and sub-species of freshwater fish in the lake and its tributaries. More than half of these animals have been found in the Siluridae family (5 different species), while the rest are found in the Cobitidae family (6 different species) (Shimizu et al., 2013).

1.5 HUMAN RELATED ISSUES AND CONSERVATION

Of the estimated 13-15,000 species of freshwater fish stock worldwide, approximately 40-45 percent of the overall varieties are deemed to be of considerable value for the survival of the diversity of fish species. Thus, the fish living in freshwater environments comprise around. Approximately 25% of all living vertebrate organisms (approximately 55,000 described species) and 15% or less than 17% of all internationally accepted freshwater species.

Freshwater habitats are now considered to be the most endangered species on the planet. Freshwater fish are extremely vulnerable to climate changes, suggesting that fish and freshwater services that human beings use and have had a significant influence on their use and power. Asia is home to around half of the worlds human population and has a major impact on inland rivers and freshwater fish populations. On the other hand, human effects are typically less marked but also rising in areas such as the Amazon and Congo basins with a lower population density (Bert, 2007).

As we have overfished some places, some have become extinct. Our fish population is influenced by the transition to the flow of fish. New species of fish now interfere with the climate. We've also got to deal with climate change. It can also be defined as a global eutrophic phenomenon

Freshwater fish are important and well enjoyed for food, sports and ornamental purposes. The utilization of specialized fishing gear is the source of over-exploitation and the extinction of many fish species worldwide as a result of increased fishing. Beyond that, the diversity of fish in the ocean is also at risk of limited pesticide-based harvesting, electrofishing, dynamite, etc.

Such major hurdles to the expansion of the water access network include diversions, extraction and storage, such as the daguerreotype regulation of the power of the irrigation dam. The extinction of wild species is due to the depletion of habitats of the surrounding oceans. A large number of dams in the world, most constructed since the end of the industrial revolution and due to insufficient land management, has resulted in biodiversity destruction, flora and fauna destruction, flood control value, the relocation of natural floodplains and the substitution of natural river formation.

Over-concentration of nutrients in the freshwater environment has been a significant indirect transformation element in the ecosystem in recent decades. Around 1960 and 2003, the global consumption of nitrogen fertilizers grew almost eightfold from 10.8 million to 85.1 million tonnes. The most common problem that affects lake/reservoir water is the question of eutrophication. As a result of eutrophication, fish populations within the ecosystem of the system are altered and the distribution of fish stocks is modified (Bert, 2007).

New species (for several reasons, including aquaculture and fishing bloating, drift nets, commercial use of wild fish, etc.) have been introduced into the world's ecosystem, including deliberate and accidental releases of aquarium species, pest-weed habitat protection, etc. Suppose fish caught in the Danube River was later harvested in Europe in Rome. In certain parts of the Black Sea, south and west of Athens, in Greece, fish are captured and harvested from the Hellespont, commonly known as the Bosporus Strait (the Stroe Waterway). Historical evidence indicates that various freshwater fish types were likely to be moved from one lake to another during the Roman Naval period and in the Middle Ages of Europe. Many scientists suggest the carp may only be found in the rivers and streams of its natural ecosystems, and that other animals may have been used for carp transport, as was the case with tilapias. For the past century, these fish have been effectively introduced to new locations.

The waters of the world have shifted with the arrival of non-indigenous aquatic plants and animals and the loss of natural species, making it more difficult for coastal habitats to separate themselves from freshwater environments. In the United States, there are much more species of plants that have been introduced since the Europeans settled down on the continent than when the immigrants arrived. With all the known organisms that still exist, we can conclude that our overall estimate is 25.2. You cannot locate a single organism that has a higher distribution than 25.2 (Lawrence & Hemingway, 2008).

While the degree of diversity of freshwater plants has undoubtedly not been extinguished, certain areas of the world have an ecology that is under pressure and, in some cases, at risk of extinction. In just a few years, hundreds of freshwater fish are close to dying out. Some of the very large tropical lakes in Eastern Africa, such as Lake Victoria and Lake Baikal, have a very high and special heritage significance. Both living beings operate as natural labs, and conservation plans should be given high priority.

The sustainability of the diversity of freshwater fish is vital to the quality of the use of water and the quality of the surrounding land. In the case of movement of marine fauna, though, the authority will be required to protect the watersheds upstream of the drainage canal, as well as the adjacent environments, the riparian areas and downstream. However, scarcely any other conditions have ever been fulfilled.

CHAPTER 2

MOLECULAR GENETICS

CONTENTS

2.1 INTRODUCTION

Every species of living entity has an implicit rule that obtains a specific group of inherited features that distinguishes it from other species. Each species requires its own developmental plan that is encoded in the DNA molecules pre-sent in its cells, often defined as a form of "blueprint" for developing the organism. The features that are acquired are determined by this developmental strategy. Since organisms follow the same evolutionary plan in the same species, entities that are members of the same species typically replicate one another, while variations between males and females are typically some significant exceptions. For starters, a human being can easily be differentiated from a monkey or a gorilla. Typically, a human being stands erect and has longer legs, relatively short hair on the body, a large brain, and a straight face with a large nose, jutting chin, separate lips, and small teeth. Many of these features, part of our evolutionary strategy, are inherited and help to set us apart as Homo sapiens. Yet human beings are not alike by all means. Most traits, or measurable features, vary from one person to the next. There is a good deal of difference in hair color, eye color, color of the face, height, weight, traits of personality, and other features. Biologically, certain human traits are transferred, others culturally. Biological inheritance comes from the color of our skin, but cultural inheritance comes from the native language we learnt as an infant. Biological descent and environmental influences collectively affect several traits. Weight, for instance, is influenced by ancestry in part, but often by the environment in part: however much food we consume, its nutritional value, our exercise schedule, and so on. Genetics is the analysis of traits that are biologically inherited, plus traits that are partially affected by the environment (Beaumont &Hoare, 2008).

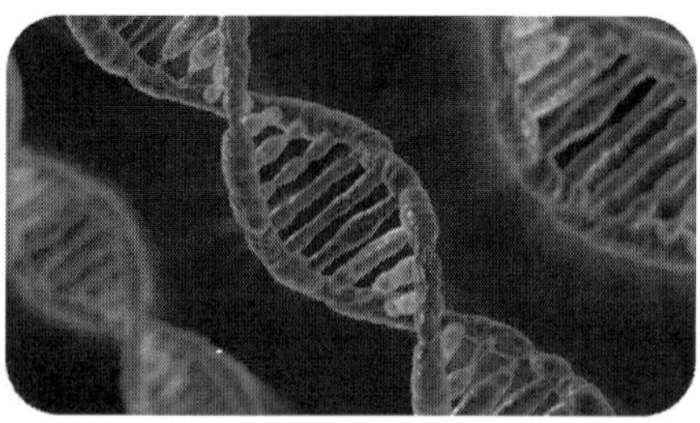

Figure 2.1: Nowadays, the genetic material is generally referred to be DNA. Scientists realized that these molecules were around for a long time. You realized that in organic molecules genetic material had been present.

Source: https://www.ck12.org/book/ck-12-biology/section/7.0/

The key notion of genetics is that: hereditary traits are determined by the hereditary aspects transmitted by parents through descendants in procreation, which are called genes in their genetic heritage. Gregor Mendel first articulated the existence of genes and the rules for the transition in 1866. Various elements of parents to children (he called them "factors") were represented in the Mendel inheritance formulation. His subjects of research were pea gardens with different features, such as pea color and plant length. At one point genetics could be studied only by the progeny produced by matings. Genetic differences between species were difficult to detect since organisms of different species usually do not match or produce sterile or die-centered hybrid offspring. This approach is also called classical genetic engineering for genetic research, organismic or morphological genetics. Due to molecular or modern genetic advances, contrast and DNA analysis may be used for the distinction between species. Classical and molecular genetics are not inherently separated. They are separate and complementary modes of studying the same work: the genetic material's work. We provide several examples in this book that demonstrate how to use molecular and classical genetics in conjunction to improve the power of genetic analysis.

Three years following Mendel's results were published, the genetics system as a molecular science was completed by 1869. Friedrich Miescher described a new type of weak acid in the nuclei of the white blood cells in 1869. Miescher's weak acid was the chemical compound now called DNA (deoxyribonucleic acid). For several years, the biological role of DNA had been uncertain and its heritage function had been unknown. This first section describes how DNA was finally isolated and determined as the material from which genes are formed (Beaumont &Hoare, 2008).

Molecular genetics is a sub-field of biology that investigates how variations between species occur in the configurations or expression of DNA molecules. In order to identify the structure and/or role of genes in the genome of an entity utilizing genetic screens, molecular genetics also applies an "investigatory approach." The field of research is focused on the synthesis of many biological sub-fields: conventional Mendelian heritage, cell biology, molecular biology, biochemistry, and biotechnology. To connect a gene sequence to a particular phenotype, researchers search for mutations in a gene or induce mutations in a gene. Molecular genetics is a strong technique for connecting mutations to genetic disorders that can assist in the quest for different genetic disorder treatments/cures. The data gathered by early 20th century scientists offered conclusive proof that the carriers of genes are chromosomes. But as did the process by which they exercise their impact, the existence of the genes themselves stayed a mystery. These basic questions were answered by molecular genetics, an analysis of the function and structure of genes at the molecular level.

In DNA molecules, genes are repetitive chains of nucleotides. A double chain of nucleotides comprises of each DNA molecule. DNA consists of four groups of nucleotides: adenine, thymine, cytosine, and guanine. In the shape of a double helix, a couple of nucleotide chains in a DNA molecule curl around each other. Hydrogen bonds among nucleotides from neighboring chains connect the two chains in the helix. Hydrogen bonding is unique such that the guanine in each chain is indeed placed in the neighboring chain next to cytosine (and vice versa) and thymine is always placed beside adenine in each chain (and vice-versa). Therefore, in a DNA molecule, the linear sequences of nucleotides in each nucleotide chain is complementary to the linear sequence of nucleotides in the remaining DNA molecule chain. A gene in any of the clusters of a DNA molecule is a section of nucleotides. For instance, not all strings of nucleotides in DNA are a gene; as per what they do, fragments of DNA are known as genes.

How do genes replicate? The definition that genes are fragments in a double helix of DNA gives a clear solution to this issue. As the coupled chains of a DNA molecule unwind, genes are diligently repeated and fresh chains are generated by the combining of complementary nucleotides alongside the splitting strands. Two versions of the initial double helix have been created when the cycle is completed, and hence the genes in the initial DNA molecule have been reproduced successfully.

What is the function of a gene? Relatively speaking, in the production of RNA molecules, genes act as models. The consequence is that the linear nucleotide sequences in a freshly synthesized RNA molecule correlate to the linear nucleotide sequences utilized as the template in the DNA fragment. Different RNA molecules perform various functional tasks in the cell, and in the production of polypeptide molecules, several RNA molecules serve a template role. Linear amino acid chains that comprise proteins are freshly synthesized polypeptides, and proteins perform a broad range of specific roles in cells and organisms (and environment). A polypeptide's capacity to act in particular ways relies on the linear sequence of amino acids that it is made of. So this linear sequence correlates to the linear sequencing of nucleotide triplets in RNA (codons), and, in exchange, correlates to the linear sequencing of nucleotides in DNA fragments, and the gene for that polypeptide is this latter section (Solé-Cava et al., 2013).

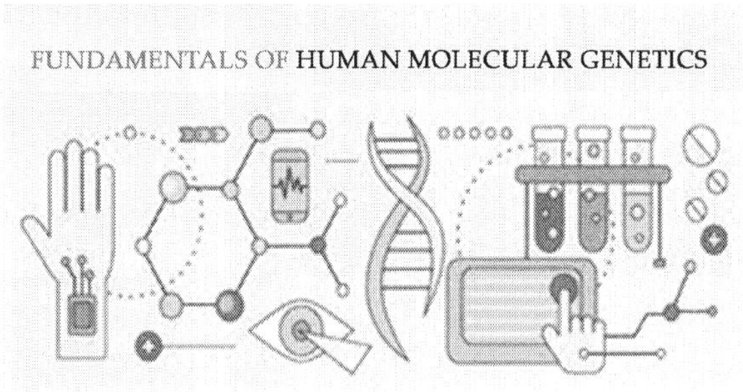

Figure 2.2: Since we could sequence the human genome as a whole, the study of human genetics has changed drastically. Researchers may now use a range of tools to classify such genetic variants that affect our health.

Source: https://courses.independent.co.uk/training/global-edulink/fundamentals-of-human-molecular-genetics-642807

How can variations in genes bring forth variations in phenotypic traits? The reasonable answer to the query 'What do genes perform?' given above is It offers the basis for clarifying how gene variations can lead to disparities in phenotypic traits. A variation in the nucleotide sequence of a gene will lead to a variation in the nucleotide sequence of RNA molecules, resulting in a change in the polypeptide amino acid sequence. Variations in polypeptide linear sequences of amino acids (and in operational RNA molecules in linear nucleotide sequence) can influence the functions they perform in the cell and organism, often having an impact that can be seen as a phenotypic variation. Mutations (gene variations) detected by the Morgan group (e.g., the purple mutation) were regularly reported as nucleotide sequence variations in DNA (Antonio M. Solé-Cava, Claudia A.M. Russo, & John P. Thorpe, 2013).

2.2 HISTORY

Several new findings were important for molecular genetics to evolve as a discipline. To define the molecule responsible for heredity, the discovery of DNA as a means to pass the genetic code of life from one cell to the next and across generations was important. The isolation of endonuclease restriction in *E. coli* by Arber and Linn opened up the genetic engineering sector in 1969. Restriction enzymes were utilized to linearize DNA for electrophoresis isolation, and Southern blotting was used to identify specific segments of DNA using hybridization probes.

In 1971, the initial recombinant DNA molecule and the initial recombinant DNA plasmid were developed by Berg using restriction enzymes. In 1972, the first recombinant DNA organism was developed by Cohen and Boyer by injecting recombinant DNA plasmids into E. Coli, now called the transformation of bacteria, opened the path for molecular cloning. In the late 1970s, initially by Maxam and Gilbert, and then by Frederick Sanger, the production of DNA sequencing techniques was crucial to molecular genetic analysis and allowed scientists to start conducting genetic screens to link genotypic sequences to phenotypes. Created by Mullis (1985), polymerase chain reaction utilizing Taq polymerase allowed scientists to produce millions of copies of a DNA sequence that would be utilized or controlled through agarose gel separation for transformation. The first whole genome (*Haemophilus influenzae*) was sequenced a decade afterward, followed in 2001 by the ultimate sequencing of the human genome through the Human Genome Project.

A new field termed genomics, which ties the molecular framework of a gene to the protein or RNA encoded by that DNA segment and the operational expression of that protein within an organism, was the result of all of those discoveries. Today, genomics is studied in many model organisms via the use of molecular genetic techniques and data is being gathered in computer databases such as NCBI and Ensemble. Bioinformatics is the computer analysis and analysis of genes within and between different organisms, and connects genetic mutations on an ecological scale (Overturf, 2009).

2.3 DNA AS THE HEREDITY REGULATOR

Johann Friedrich Miescher, a Swiss chemist, extracted a material comprising phosphorus and nitrogen from cell nuclei in 1869. Initially, this material was named nuclein, but now it is recognized as deoxyribonucleic acid, or DNA. DNA is the chemical portion of chromosomes which, in microscopic formulations, is primarily accountable for their staining effects. Because eukaryote chromosomes include a number of proteins in addition to DNA, the issue inevitably arose as to if the carriers of genetic information were nucleic acids or proteins, or all. Until the beginning of the 1950s, many biologists were compelled to assume that proteins were the main inherited carriers. There are only four distinct unitary bases of nucleic acids, but 20 different amino acids are built up of proteins. Therefore, proteins seemed to have a larger variety of structures, and at first, the variety of the genes appeared likely to rest on the heterogeneity of the proteins.

In exquisitely straightforward microbiological experiments, proof that DNA functions as the bearer of genetic data were first strongly illustrated. Two strains of the bacterium *Streptococcus pneumoniae* were examined in 1928 by the English bacteriologist Frederick Griffith; a strain was fatal to mice (virulent) another was safe (avirulent). Griffith observed that both the heat-killed virulent bacteria or the living avirulent bacteria-infected mice stayed infection-free, but both mice infected with a combination were sick and perished. It appeared as if some chemical "transforming principle" had moved and modified the avirulent cells from the dead virulent cells. In 1944, Oswald T. Avery, an American bacteriologist, and his colleagues found that DNA was the transformative element. Avery and his research team collected blends from heat-killed virulent bacteria and also RNA (ribonucleic acid, a similar molecular equivalent of DNA), DNA, lipids, proteins or polysaccharides (sugar subunits) were inactivated and avirulent cells were individually applied to each form of a mixture. The only molecular class where the transition to virulence was stopped by inhibition was DNA. It therefore appeared that the hereditary content had to be DNA, so it could change (Beaumont &Hoare, 2008).

The analysis of bacteriophages, viruses that invade and destroy bacterial cells, has produced a similar hypothesis. Hundreds of bacteriophage offspring are formed from a host cell contaminated by one bacteriophage. Two species of bacteriophage samples were collected by American biologists Chase & Hershey, 1952. The external protein coat of the bacteriophages was marked with a radioactive isotope in one population; the DNA was marked in the other. After encouraging either species to invade bacteria, Hershey and Chase noticed that the progeny bacteriophage contained radioactivity only after DNA was labeled. They then hypothesized that DNA is inserted into the bacterial cell, where, at the cost of the host, it guides the production of multiple complete bacteriophages. In other terms, the hereditary material liable for the essential traits of the virus is DNA in bacteriophages.

Utilizing externally added DNA, in a way identical to that utilized by Avery for bacteria, the genetic makeup of many species can be converted today. DNA transformation is capable of moving through cellular and nuclear membranes and then integrating into the receiver cell's chromosomal DNA. Moreover, utilizing modern DNA technology, it is feasible to separate the segment of chromosomal DNA that comprises a specific gene, modify its composition, and reintroduce it into a cell to cause changes that prove beyond question that the DNA is liable for a large part of the general traits of an organism. For reasons like these, it is now known that genes are made of DNA in all living species, except for certain viruses (Beaumont &Hoare, 2008).

2.4 COMPOSITION AND STRUCTURE OF DNA

Many concerns were still unresolved by the assumption that DNA is the genetic material. Whenever a cell divides, how is the DNA in a gene recreated? How does an inherited trait regulate the DNA in a gene? When a mutation (a shift in DNA) occurs in a gene, what occurs to the DNA? In the early 1950s, in the expectation that the composition alone would provide solutions to these concerns, a variety of researchers started trying to explain the detailed molecular structure of DNA. The first fundamentally accurate three-dimensional design of the DNA molecule was suggested in 1953 by James Watson and Francis Crick at Cambridge University. In its beauty, the design was dazzling and groundbreaking in proposing how DNA duplicates itself, regulates genetic features, and undergoes mutation. Although their DNA molecule tin-and-wire model was still unfinished, Crick would frequent his best pub and announce, "We have found the secret of life."

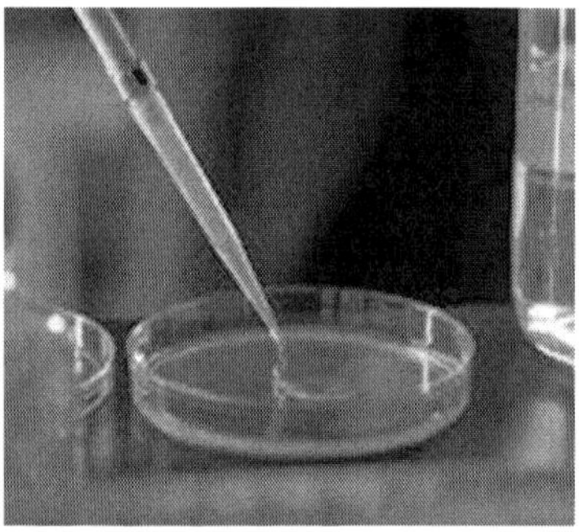

Figure 2.3: Classical genetics are classified as techniques and methods of genetics. Genetics contributes to molecular biology. The discovery of genetic codes and cloning devices, such as restrictive enzymes, opens up many new genetic fields.

Source: https://hsc.unm.edu/medicine/departments/genetics-microbiology/

DNA comprises of two long chains of subunits in the Watson-Crick system, each twisted over the second to form a double-stranded helix. It is a right-handed double helix, which suggests that each chain assumes a clock-wise direction as it moves as one looks down the barrel. If you picture yourself staring up from the base into the framework, you can envision the right-handed coil. The dark spheres represent each specific strand's "backbone," and they curl in a clockwise fashion. The subunits within every strand are nucleotides, each of which comprises all of the four chemical components known as bases bound to the 5-carbon sugar deoxyribose phosphorylated molecule. In DNA, the four bases are Cytosine, Thymine, Guanine, and Adenine (Overturf, 2009).

If one strand contains an Adenine, then the companion strand has a Thymine; and if one strand contains a Guanine, then the companion strand contains a Cytosine, at any location on the paired strands of a DNA molecule. It is stated to be similar to the combination of Adenine and Thymine as well as of Guanine and Cytosine; Thymine is the counterpart of Adenine and Cytosine is the counterpart of Guanine. The additional matching implies that on the opposite strand, each base on one strand of the DNA is paired with a base in the opposite location. In addition, nothing limits the sequence of bases in a single strand, so that along one strand any sequence may be present. This theory describes how only four DNA bases can code for the enormous amount of information necessary for an organism to be formed. It is the series of bases along the DNA that encodes the genetic material, and there is no restriction on the sequence.

The corresponding combination has also labeled the pairing of Watson-Crick. The base pairs are described in the three-dimensional framework by the smaller spheres occupying the double helix's inside. Like shillings in a roll, the base pair lie flat, placed on top of each other perpendicular to the long axis of the double helix. Biologists sometimes refer to specific strands as single-stranded DNA and to the double helix as double-stranded DNA or duplex DNA when describing a DNA molecule.

Every DNA strand contains directions or polarity, just like a cord of circus elephants attached trunk to tail. Any other elephant in this analogy corresponds to one nucleotide in the DNA chain. By the way, the nucleotides are indicated, the polarity is discouraged. The 3' end of the strand is marked with the "trunk" end, and the 5' end is marked with the "tail" end.

Knowing the structure of DNA immediately gave clues to its work beyond the most optimistic hopes:

- In a series of bases, DNA could include genetic data in coded format, similar to letters written on a piece of paper.
- Alterations in genetic data (mutations) can occur from duplication errors in which the DNA base sequence has been changed.

- The DNA sequence of bases could be repeated by using each of the various 'partner' strands as just a prototype for creating a new partner strain that includes model strands of bases.

2.5 PROTEINS AND GENES

Now that we have an obvious understanding of the functional structure of the genetic plan, how can the evolutionary strategy become a complex living organism? The genes are composed of distinctive words which create paragraphs and phrases that make sense to the letter design when considered as a string of letters on a sheet of paper. Protein, a macromolecular class that performs over all the functions in the cell, is derived from complex and complex DNA codes. The structural proteins providing cell firmness and versatility, proteins forming cell membrane pores to regulate the passage into or out of cells of small molecules and receptor proteins regulating cell reaction activity in the developing media or other cells are mainly proteins.

In many of the metabolic processes of cells, proteins are also essential. They are important for the production and breakup of organic molecules and the production of the chemical energy necessary for cellular activity. The word enzyme was developed in 1878 to relate to biological catalysts that speed up cellular biochemical reactions. Centered mainly on the work of German biochemist Emil Fischer, enzymes were seen as protein by 1900. The "errors" of nature often indicate how things work, as happens in science. This has been the case with the creation of the relationship between genes and disease, since a "mistake" in a gene (mutation) could lead to a "mistake" in the protein involved. This offered a fruitful mechanism of research in the analysis of genetics (Solé-Cava et al., 2013).

2.6 INBORN METABOLISM ERRORS AS A SOURCE OF GENETIC DISEASE

At the turn of the 20th century, English physicist Archibald Garrod found that some heritage diseases followed the principles of propagation developed by Mendel for his garden peas. In 1908 Garrod lectured on the relation between heritage, enzymes and disease as a fundamental theory: the heritage deficit in an enzyme contributes to all genetic disorders where the cellular metabolism is abnormal. These disruptions are known as innate metabolism mistakes, a term still used today. Garrod studied multiple innate metabolic agents in which patients excrete irregular compounds in the urine.

One of these was Alkaptonuria. The abnormal substance is extracted in this case, homogentisic acid.

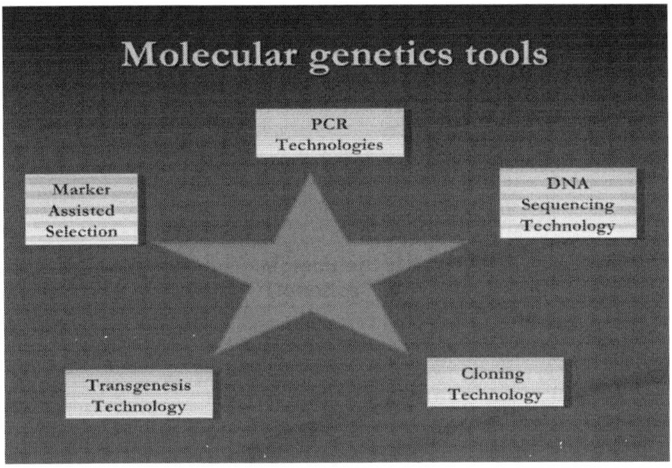

Figure 2.4: Molecular genetics relies on molecular gene structure and function. Molecular genetics was used both in traditional and molecular genetic engineering (e.g. hybridization).

Source: https://nowbotjourney.netlify.app/molecular-genetics.html

Alkapton, hence the term alkaptonuria for the disorder, was an earlier name for homogentisic acid. While alkaptonuria is rare was well established even before Garrod studied it, with an occurrence of about one in 200,000 individuals. The disorder itself is comparatively benign, but has one striking symptom: due to the oxidation of homogentisic acid, the patient's urine turns black. This is why black urine disorder is often referred to as alkaptonuria. Garrod was primarily engaged in the biochemistry of alkaptonuria, but family studies revealed that the disorder was inherited as if because of a certain gene defect. With regard to biochemistry, he concluded that alkaptonuria did not overcome the six-carbon phenyl ring in homogentisic acid in a patient. Where's this ring going through? It was purchased from food in its diet by most species. Garrod indicated that homogentisic acid originally comes in two amino acids, tyrosine and phenylalanine, which both have a phenyl ring, as a degradation product. One of the "building blocks" from which proteins are produced is an amino acid. The components of ordinary proteins include tyrosine and phenylalanine. A biochemical pathway or a metabolic pathway is the name of any other series of biochemical reactions. The transformation from the "input" or substrate molecule, to the "output" or product molecule, representing a single phase. He actually understood, however, that the splitting open of the phenyl ring is the crucial step in the degradation of homogentisic acid and that the phenyl ring in homogentisic acid arises from dietary tyrosine and phenylalanine (Solé-Cava et al., 2013).

What makes a biochemical pathway possible for each step? Garrod accurately assumed that every step of the process would involve a specific enzyme. Persons with an inherent metabolic error have a single metabolic error, like alkaptonuria, since that stage has not a working enzyme. At that point, when an enzyme is defective, it is stated to have a barrier. One common consequence of a disrupted pathway is that the faulty enzyme's substrate accumulates. One common consequence of a disrupted pathway is that the faulty enzyme's substrate accumulates. Garrod has shown that an enzyme whose function is to open the homogentisic acid ring should be present and that the enzyme is absent for these patients by monitoring the aggregation of homogenetic acid in patients who suffer from Alkaptonuria. At present, isolation of the homogentisic acid phenyl ring enzyme was not carried out until 50 years after lectures by Garrod. The enzyme, as Garrod had predicted, is deficient in people with alkaptonuria in liver cells in normal people.

The mechanism for phenylalanine and tyrosine degradation is relevant as it is currently understood. The emphasis is on the enzymes and not on the metabolite systems or small substances on which the enzymes act. The existence of a specific enzyme the catalyzes that step needs each step in the pathway. Although the alkaptonuria in which 1,2 dioxygenase is home to the defective enzyme was only known to Garrod, we now recognize the clinical consequences of other enzyme defects. The others are rather dangerous, unlike the comparatively mild legacy alkaptonuria. Phenylketonuria (PKU) is caused by the lack or a defect in the phenylalanine hydroxylase enzyme (PAH). When this phase in the road is stopped, phenylalanine rises. Excess phenylalanine is broken down into dangerous metabolites causing myelin formation defects to weaken a child's developing nervous system and cause severe mental retardation (Solé-Cava et al., 2013).

However, if PKUs are detected early after being born in infants, they can be put on a specially formulated diet that is low in phenylalanine. Only as much phenylalanine as can be utilized in protein synthesis is permitted in the infant, so excess phenylalanine does not aggregate. It's very strict about the special diet. Milk products, milk, eggs, fish, poultry, and meat, legumes, nuts and bakery products made with standard flour are exempt. An expensive synthetic solution replaces these foods. However, the adverse effects of excess phenylalanine on mental growth can generally be prevented with a special diet, even though the fetus is at risk in adult women with PKU who are expecting.

In several nations, including the United States, all newborns get their blood checked for PKU chemical symptoms. Since PKU is relatively normal, regular screening is cost-effective. In the United States, among Caucasian births, the incidence is around 1 in 8000. In other ethnic groups, the disease is less prevalent. In the metabolic pathway, forms of tyrosinemia are caused by defects in the degradation of tyrosine or 4-hydroxyphenylpyruvic acid. There are serious illnesses as well. Type II is related to mental retardation and skin lesions; type III is correlated with extreme liver dysfunction (King, 2013).

2.7 DEFECTIVE PROTEINS AND MUTANT GENES

It means that a defective enzyme comes from the function of Garrod from a mutated gene, but how? The speculations were not from Garrod. For all that he knew, genes were enzymes. This may have been a logical theory at that time. The relation between genes and enzymes is now understood to be somewhat dependent. With few exceptions, each enzyme is encoded in a particular sequence of nucleotides in the DNA region. The "gene" in which the enzyme is encoded is comprised of the enzyme-coding DNA region and neighboring regions regulating when or in which cells the enzyme is produced. The standard typographic standard is used in the list below, whereby genes are not printed in italics, for italics. It is useful because the protein portion of a gene can be described by the same symbol as the gene themselves but the protein symbol is not, while the protein symbol is in italics:

- The PAH gene on chromosome 12's long arm encodes phenylalanine hydroxylase (PAH).
- Tyrosine aminotransferase is encoded by the TAT gene on the long arm of chromosome 16 (TAT).
- The HPD gene encodes 4-hydroxyphenylpyruvic acid dioxygenase on the long arm of chromosome 12 (HPD).

2.8 THE ENVIRONMENT AND GENES

The general theory that genes code for proteins and that mutant genes code for mutant proteins is demonstrated by innate metabolism errors. Mutant proteins undergo such a dramatic change in metabolism in cases such as PKU that a significant genetic deficiency results. Yet biology does not actually constitute fate. The climate also influences species. PKU acts as an illustration of this concept since, within the normal range, patients who conform to a diet limited by the amount of phenylalanine improve mental capacity. In this case, what is true is generally true. The interaction between genes and the environment influences most traits. It is indeed true that many genes influence most traits. No one understands how many genes are implicated in the brain and nervous system's growth and development, however, the number has to be in the thousands. This amount is in addition to the genes that are needed to perform metabolism and other essential life functions in all cells. When contemplating severe cases, such as PKU, in which a single mutation can have such a dramatic impact on mental growth, it is simple to lose sight of the multiplicity of genes. For every complex machine, the condition is the same as that. If thousands of components work together in unison, an airplane can work, but only one faulty part can reduce it closer if that part affects a critical system. Similarly, a large number of genes operating in harmony enable the creation and functioning of each trait, but in some instances, a single mutant gene may have disastrous consequences (Shimizu et al., 2013).

The relationship involving a gene and a trait is, in other words, not always a clear one. Organism biochemistry is a complicated web of branches in which different enzymes can exchange substrates, generate the same products, or be receptive to the same regulatory elements. The consequence is that many genes that function together and in conjunction with environmental factors are the net result of the most noticeable traits of species. Examples of each of the three concepts governing these interactions are given by PKU:

- More than one characteristic may be caused by one gene. Blond hair and decreased body pigment are frequently observed in children with severe types of PKU. This is since the lack of PAH, the precursor of the pigment melanin, is a metabolic block that inhibits the conversion of phenylalanine into tyrosine. Only if one understands the metabolic interactions between phenylalanine, tyrosine, and melanin does the association between extreme mental retardation and reduced pigmentation in PKU make sense. The characteristics would appear entirely unrelated if these similarities were not known. In this respect, PKU is not uncommon. Via their alternative or indirect effects, many mutant genes affect several traits. Pleiotropic effects are called the different, often apparently unrelated, effects of a mutated gene, and the condition itself is known as pleiotropy. A cat having white fur and blue eyes, a trend of pigmentation that is frequently associated with deafness (about 40 percent of the time). Deafness can also be seen as a pleiotropic consequence of the hue of the white coat and blue eyes. There is no understanding of the developmental basis of this pleiotropy.

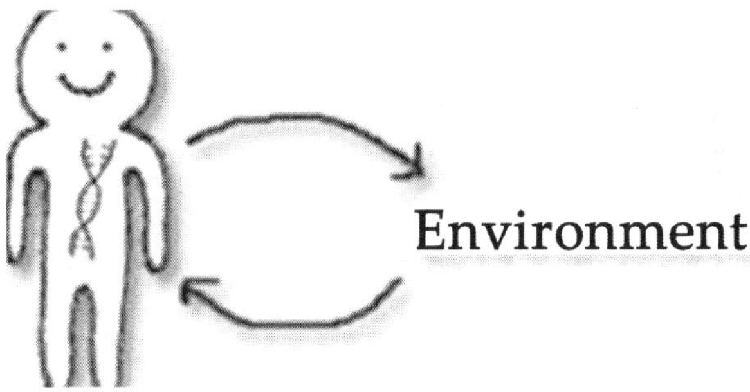

Figure 2.5: A individual may respond or act against his/her external environment on the basis that certain acts in a given environment are considered acceptable or unacceptable. Aggression in turn may be triggered by issues with the family, social pressure, job tension and other concerns.

Source: https://academichelp.net/samples/academics/essays/analysis/genes-and-environment.html

- More than one gene can influence any phenotype. In conjunction with the huge amount of genes that are needed for the normal growth and functioning of the brain and nervous system, we addressed this theory earlier. Amongst these are genes that influence the blood-brain barrier structure, which comprises specific glial cells bundled in the brain around tight capillary walls, creating an obstruction to the passage from the blood to the brain of most water-soluble molecules. Therefore, the blood-brain barrier influences the degree to which excessive free phenylalanine in the blood can penetrate the brain itself. PKU patients with very identical blood phenylalanine levels may have significantly varying levels of cognitive function because the efficacy of the blood-brain barrier varies among individuals. This also demonstrates in part why compliance to a regulated diet of phenylalanine in children is extremely significant, but less so in adults; in children, the blood-brain barrier is less well-formed and thus less efficient in blocking excess phenylalanine. Also, simpler metabolic characteristics are affected by several genes. Phenylalanine decomposition and excretion help as a convenient example. As variations in the activity of each one of them can affect the decline and excretion rate of phenylalanine, it is important for all pathways enzymes to determine the excessive amount of phenylalanine in PKU patients' blood.

- Many characteristics are affected by both environmental and genetic factors. Here we return to the diet for low-phenylalanine. PKU children may not have a significant medical disorder. Their capability can be integrated into the average limits by nutritional therapy. PKU shows what genetics was motivated to work for the discovery of the molecular base of the inherited condition. It is expected that knowing the metabolic cause of the disease will eventually allow the severity of the condition to be eased through diet, medication or other therapies.

2.9 CENTRAL DOGMA

In the analysis of molecular genetics, the Central Dogma is the foundation of all genetics and assumes a crucial role. DNA replicates itself, DNA is transcribed into RNA, and RNA is converted into proteins, the Central Dogma says. The genetic code is utilized along with the Central Dogma in determining how RNA is converted into proteins. DNA transcription and replication from DNA to mRNA happens in the mitochondria, whereas in the ribosome there is translation from RNA to proteins. The genetic code is composed of four base pairs: guanine, uracil, cytosine, and adenine, and is repetitive, meaning that the similar amino acid is formed by several combinations of these nucleotides (read in triplicate). Proteomics and genomics are areas that emerge from the analysis of molecular genetics and the Central Dogma of biology.

In suggesting that the genetic data in DNA is found in the sequence of bases in a way similar to letters written on a strip of paper, Watson and Crick were right. The genetic code for the protein is found in only one strand in a region of DNA that guides the production of a protein, and is decrypted in a linear order. One or more polypeptides chains form a standard protein; each polypeptide chain comprises a linear series of linked end-to-end amino acids. For instance, four similar polypeptide chains comprise the enzyme PAH, all 452 amino acids in length. Each subsequent "code word" in the DNA in the decoding of DNA determines the following amino acid to be introduced as it is being introduced to the polypeptide chain. The whole gene is much longer—about 90,000 pairs of nucleotides. Just 1.5 percent of the gene is dedicated to amino acid coding. Some sequences that regulate the function of the gene are included in the noncoding portion, but how much of the gene is engaged in regulation is not known (Dunham, 2011).

There are 20 distinct amino acids. Just four bases code for the 20 amino acids, each of which contains three neighboring bases and the "word" of the genetic code. For eg, the basic ATG series specifies the methionine (Met) amino acids, the TCC serine (Ser) and the ACT threonine (Thr) and the GCG alanine (Ala). There are 64 possible three-base combinations, but only 20 amino acids, since some variations code with the same amino acid. DNA duplex is the human series of codes in the PAH polypeptide chain for the initial seven amino acids. Protein DNA codes, not directly but indirectly, through transcription and translation mechanisms. As the basic dogma of molecular genetics, the alternative route of knowledge transmission is established. The term dogma is a "set of beliefs" which dates from the first time the principle was introduced as a theory. Since then the «dogma», but the phrase remains, has been checked experimentally (Dunham, 2011).

The primary principle is that DNA does not specifically code for protein, but instead operates through a ribonucleic acid intermediary molecule. The design of RNA is similar to that of DNA, but not identical. There is a sugar discrepancy (RNA produces sugar ribose rather than deoxyribose), RNA is typically single-stranded, and RNA includes base uracil (U) instead of DNA-based thymine (T). Three forms of RNA specifically take part in the synthesis of proteins:

- A messenger RNA (mRNA) molecule that bears genetic data from DNA and is utilized as a polypeptide synthesis template. There is a high proportion of nucleotides in many mRNA molecules that specifically code for amino acids. For instance, the size of the mRNA for PAH polypeptide codes for 452 amino acids in 2400 nucleotides; in this instance, greater than 50 percent of the size of the amino acid mRNA codes.

- Various types of ribosomal RNAs (rRNA) constitute major ribosomal materials known as cellular components on which polypeptide is formed.
- A number of RNA (tRNA), each of which is distinguished both by a particular amino acid and a three-base identifying area that pairs the mRNA base with three adjacent bases. As each tRNA contributes to translations, the amino acid reflects the terminal subunit connected to the length of the growing polypeptide chain. TRNA is referred to as tRNAMet, tRNASer is referred to as serene carriers, etc.

The basic theory of molecular genetics is the central dogma, since it illustrates how the genetic data in DNA is represented in a polypeptide chain in the amino acid sequence: The nucleotide sequence in the gene determines the nucleotide sequence in the messenger RNA molecule; the nucleotide sequence in the messenger RNA in effect determines the amino acid sequence in the polypeptide chain.

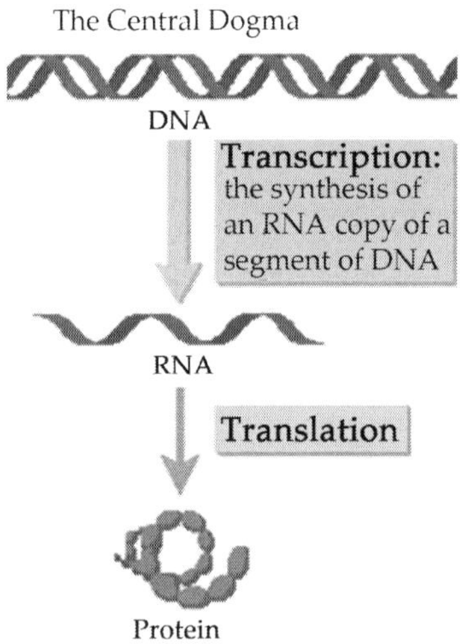

Figure 2.6: The 'Central Dogma' is the mechanism by which the DNA instructions are translated into a working product. In 1958, it was first suggested by the DNA structure revealer Francis Crick.

Source: http://www.phschool.com/science/biology_place/biocoach/transcription/overview.html

What may account for the increased complexity of RNA intermediaries, provided a system as conceptually easy as protein DNA coding? One potential explanation is that, for instance, by degrading the mRNA for an unneeded protein, an RNA intermediate provides another degree of power. History may be another potential explanation. The design of the RNA is unusual in having both a data content present in its base sequence and a dynamic, folded three-dimensional structure that gives catalytic activities to certain RNA molecules. Many scientists agree that RNA was used for both genetic knowledge and catalysis in the early forms of life. The informative role was moved to DNA and the catalytic role to protein as evolution progressed. However, as a go-between in the transmitting information and protein synthesis processes, RNA has been locked into its central position.

This hypothesis suggests that RNA's involvement in protein synthesis is a remnant of a "molecular fossil" from the earliest stages of evolution. A number of findings support the hypothesis. For instance, in order to get underway, (1) DNA replication needs an RNA molecule, (2) an RNA molecule is important in the production of the points of the chromosomes, and (3) certain RNA molecules function to catalyze key protein synthesis processes (Caliskan et al., 2015).

2.10 ADVANTAGES OF DIAGNOSES BASED ON DNA

The detection of Mendelian genetic abnormalities has so far only been limited to cases of biochemical alteration in fetal tissue and seen (e.g., blood, chorionic villi, amniotic fluid cells, skin, liver). Hemophilia A and anemia of sickle cells, procedures conferring significant risk on the fetus, can only be observed, for example, by fetal blood tests and phenylketonuria (PKU) by fetal liver biopsies. As the biochemical basis was unclear, diseases like cystic fibrosis (CF) and Duchenne muscular dystrophy (DMD) were not observable prenatally. Utilizing biochemical or morphological markers in fetal tissues, fewer than 100 of the more than 2000 identified mendelian disorders are prenatally diagnosable. Now, however, all Mendelian disorders are potentially vulnerable to prenatal diagnosis by the use of recombinant DNA technology.

2.11 TECHNIQUES

a. *Forward genetics:-* Forward genetics is a method used in molecular genetics to classify genes or genetic defects causing a certain phenotype. Random mutations are produced with mutagens (chemicals or radiation) or transposons in a genetic screen and the particular phenotype is screened for individuals. Sometimes, in the context of a selection, a secondary assay can follow mutagenesis where it is difficult to observe the desired phenotype, for instance in cell or bacteria cultures. Using an antibiotic resistance gene or a fluorescent reporter, the cells can be transformed such that the mutants with the desirable phenotype are chosen from the non-mutants. In order to decide if the phenotype derives from more than one gene, mutants displaying the phenotype of concern are isolated and a complementation experiment may be conducted. The mutant genes are then classified as dominant, recessive (showing a cognitive impairment), or epistatic (resulting in a gain of structure) (the mutant gene masks the phenotype of another gene). Finally, through sequencing, the position and precise nature of the mutation is mapped. Forward genetics is an unbiased approach that sometimes leads to many unexpected findings, but can be time-consuming and expensive. To research phenotypes arising from gene mutations, model organisms such as the fruit fly Drosophila melanogaster, the nematode worm *Caenorhabditis elegans*, and the zebrafish *Danio rerio* have been effectively used (Caliskan et al.,2015).

b. *Reverse Genetics:-* Reverse genetics is the word used for molecular genetics methods to evaluate the phenotype of a gene of interest from an intentional mutation. To infer the role of the unmutated form of the gene, the phenotype is utilized. Random or deliberate modifications to the gene of importance can be mutations. Mutations can be a nucleotide substitution-induced missense mutation, a nucleotide inclusion or deletion to trigger a frameshift mutation, or a gene segment's full addition/deletion. Deleting a specific gene causes a knockout of the gene where the gene is not activated and the consequence is a loss of functionality (e.g. knockout mice). Mis-sense mutations, known as a knockdown, can cause complete function loss or results in minimal function loss. Knockdown can also be accomplished by interfering with RNA (RNAi). Conversely, genes may be inserted into the genome of an organism (also known as a transgene) to produce a knock-in gene and result in the host acquiring function. While these methods have some inherent bias about the decision to relate a phenotype to a specific function, since the gene of interest is already known, it is far quicker in terms of development than forward genetics.

Virtually every cell comprises the full array of genes, even if a gene is not represented by a specific cell. For example, while hemoglobin-encoding genes are expressed only by reticulocytes, all cells, even amniotic fluid cells and chorionic villi, are present with DNA coding for hemoglobin. Moreover, in order to identify the DNA that codes for the mutant gene, the exact molecular or biochemical defect need not be identified. DNA-based research depends on the fact that the DNA of different individuals has almost infinite numbers of nucleotide-sequence variations. Via restriction fragment length polymorphisms (RFLPs), or RFLP analysis, these variations can be established. While most variations in sequences have no pathological significance, they may serve as markers for disease-causing mutant genes. In families with an established genetic disorder, RFLPs allow the identification of affected individuals, even if the specific deficiency in the gene (i.e. mutation) is unclear. RFLP testing may be used (e.g., sickle cell anemia) or, instead, a range of measures may be utilized for direct mutation identification in cases where the disease-causing mutation is identified (e.g., cystic fibrosis) (Kim, 2019).

DNA-based genetic disease research uses the aforementioned technologies: DNA amplification using polymerase chain reaction (PCR), hybridization visualization of cloned DNA probes, endonuclease (restriction enzyme) digestion restriction, Southern or dot blotting immobilization of DNA, and electrophoresis isolation of DNA fragments. These methods offer important tools for the diagnosis of genetic disorders, used in a number of ways.

1. *Restriction Fragment Length Polymorphisms:-* Endonucleases are limited by bacterial enzymes, which recognize and cut dual-stranded DNA in single nucleotide sequences. More than 400 restrictive enzymes have been identified. Restriction sites are referred to as the DNA sequences recognized by various restriction enzymes. Restriction sites are evenly distributed around the genome sites and can be found inside or around any single gene. However, since 95% of human DNA is not gene-coded, many limiting sites are identified by chance within the non-

coding portion of genes. A restriction fragment is considered the DNA between two restriction sites, and the scale of a restriction fragment is defined by the gap within two restriction sites. Thus, when a restriction enzyme digests DNA, it is cut into several fragments of different lengths. As nucleotide sequences differ from person to person, the frequency of restriction locations and the length of restriction-fragment sizes will vary between individuals (i.e., restriction-fragment lengths are polymorphic).

A polymorphism relates to the existence of two or more variants of a gene in a population, the lesser frequent having a frequency of at least 1%. The ABO blood types, serum transferrin, or the red-cell enzyme G6PD are classic instances of human genetic polymorphisms. The number of red-cell polymorphisms, serum, or antigen in the human genome is, nevertheless, limited to less than 50 loci, unlike RFLPs.

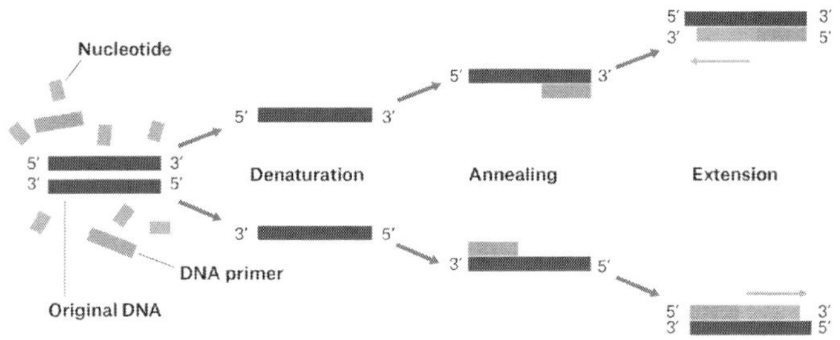

Figure 2.7: One of the most widely used biotechnologies in biology is the polymerase chain reaction (PCR). It is an important part of every student's curriculum and the majority of biomedical researchers have carried out PCR, or rely on PCR data at least. Clinical laboratories most also use PCR to assist patients in diagnosis.

Source: https://blog.labtag.com/a-brief-history-of-pcr-and-its-derivatives/?a mp

2. *Detecting RFLPs using Hybridization, Southern Blotting, and Electrophoresis :-* Fragments of different sizes exist with a particular restriction enzyme eaten with genomic DNA. Agarose gel electrophoresis on the basis of size will mechanically insult various sizes fragments (bigger restriction fragments move through the gel much slower than smaller restriction fragments). After electrophoresis, treated DNA emerges as a stain under ultraviolet light. These fragments are metabolized by a process

called "Southern blotting" (Southern blossoming) that preserves the space orientation of the restrictive pieces while immobilizing and preserving DNA in the gel and are then transferred to a membrane (like nitrocellulose or strengthened nylon). Therefore, on the membrane, the band designs are similar to any in the gel. The membrane is immersed in a solution comprising a radioactively (or enzymatically) coded, single-stranded probe (i.e., DNA that is comparable to the gene of concern or to residues that are very similar to {i.e. connected to} the gene of interest) to identify a specific gene among the several fragments on the blot. The probe can hybridize (i.e. pair) to DNA since both are single-stranded, which is complementary to it (Kim, 2019).

3. *Slot (or dot) blots:-* It is also possible to stabilize DNA by "dotting" undigested DNA directly onto a vacuum membrane. As mentioned above, these dots, or "slot," blots can be hybridized into a probe. Hybridization signals following autoradiography are perceived as a dark dot. With PCR-amplified DNA and sequence-specific oligonucleotide (SSO) probes, dot blots are most commonly used.

4. *Polymerase Chain Reaction (PCR):-* PCR is an in-vitro way of producing multiple versions of specific DNA sequences. In just a few hours, this efficient method can synthesize over one million versions of specific DNA sequences. PCR allows relatively small quantities of DNA to be diagnosed (i.e., removing the need to culture cells to produce larger amounts of DNA) and decreases the time taken from a few weeks to a few days to make a diagnosis. Diagnoses can be rendered either by direct observation of the intensified product under ultraviolet light on amplified DNA or by sequence-specific oligonucleotide probes after hybridization. A series of nucleotides that is metabolized in the laboratory is an oligonucleotide probe. The advantage of these limited sequences is that the probe can not hybridize during specific conditions unless the nucleotides in the DNA being examined match precisely the nucleotide variety in the probe. Therefore, among sequences that vary by a single base pair (e.g., sickle vs regular β-globin gene), an oligonucleotide probe can distinguish. A major drawback of PCR is that it is important to know the DNA sequence on both ends of the mutation (i.e., the flanking sequences). However, since more knowledge about disease genes overcomes this constraint, studies using PCR are becoming the tool of choice.

2.12 EVOLUTION: TO PROTEOMES FROM PROTEINS, TO GENOMES FROM GENES.

The process of excretion and breakdown of phenylalanine is by no means unique to people. Several of the interesting molecular genetics generalizations arising is that very particular organisms have many genetic and biochemical characteristics, including plants and animals. These symmetries suggest a basic "unity of life": all organisms in the world express a variety of genetic system characteristics, including genetic data encrypted in the sequence of DNA bases, RNA transcription and ribosome protein translation, with the transmission of transferred RNAs. In their biochemistry, both organisms also share many features, including several proteins, including enzymes that are identical in their sequence for amino acids, three-dimensional framework, and purpose.

DNA is considered as the organism's genome as a whole in a single cell. The genome is normally treated with DNA in a reproductive cell in sexual organizations. The human genome in the chromosomes of sperm or egg contains approximately 3 billion nucleotide pairs of DNA. The proteome is known as the complete collection of genome proteins. Genomics involves genome analysis; proteomics is proteome analysis. The unity of existence can be found in the combination of proteins in the proteomes of various organisms. The proteome of the fruit fly *Drosophila melanogaster*, for instance, comprises 13,601 proteins; these can be classified into 8065 separate protein families that are identical in the chain of amino acids. For instance, the *Caenorhabditis elegans* nematode worm proteome comprises 18,424 proteins that can be classified into 9453 families. Around 5000 proteins that are fairly similar to be considered to have a shared role share these two proteomes. In the proteome of the yeast *Saccharomyces cerevisiae*, about 3000 are also present among the protein families specific to flies and worms. It seems possible that all species whose cells have a nucleus and chromosomes will transform out to share many thousand protein families on the basis of these three fully analyzed complex genomes. In addition, about a thousand of these groups of proteins are shared with species that are as distantly linked as bacteria (Lawrence & Hemingway, 2008).

Unity of life in molecules

Why do organisms share a shared set of identical proteins and genes? Since everybody has a common source. The evolution cycle occurs as a group of species from a common ancestor gradually shifts over time. The development cycle occurs. From an evolutionary standpoint, the cohesion of fundamental molecular mechanics is established by descending from a remote group ancestor in the place of molecular mechanisms. From an evolutionary perspective, not only life's unity, but also many other characteristics of life's organisms are understandable. For instance, the occlusion of an RNA transitional in the genetic data flow from DNA to RNA into a protein is meaningful in the earliest aspects of life using RNA for both genetic and enzymatic catalysis. Evolutionary biologist Theodosius Dobjansky's popular aphorism sums up the importance of an evolutionary perspective in understanding aspects of biology that seem trivial or complicated in no unnecessary way.

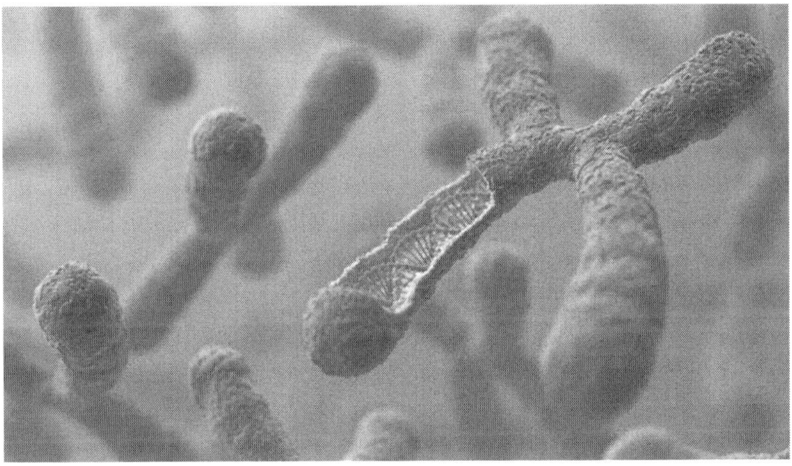

Figure 2.8: The long, ladder-like strands of DNA inside the nucleus of each cell are rolling like threads and bundled into 23 pairs of chromosomes. Each parent produces one chromosome of each pair.

Source: https://www.vision.org/outline-molecular-genetics-1120

An example is one sign of the shared heritage among the inhabitants of the Earth. The relationship tree was derived from nucleotide sequence similarity in an RNA molecule contained in the ribosome's tiny subunit. It distinguishes three main kingdoms of organisms:

- Kingdom's bacteria: This group contains the majority of cyanobacteria and bacteria (initially referred to as blue-green algae). These organisms' cells miss a nucleus and mitochondria enclosed by a membrane, are covered with the help of the cell wall, and are separated by binary fission.

- The Kingdom Archaea; this group was originally found in small organisms producing methane gas or living in insane conditions, such as elevated levels of salt, or hot springs. In more natural settings as well, they are widely disseminated. The cells of archaeans lack internal membranes, much like those of bacteria. Examination of the DNA sequence suggests that the DNA transcription and replication mechanism in archaeans match that of eukaryotes, although their metabolism is closely similar to that of bacteria. Around half of the genes present in Archaea are specific to this group of kingdoms.
- Eukarya Kingdom; This group includes all organisms with large internal membranes, mitochondria and membrane-based nuclei in their cells. DNA is divided into chromosomes and cell division takes place by mitosis. Plant and animal plants and fungi as well as many single cell species such as amoebae and ciliated protozoa are present in eucaryotes.

In contrast to eukaryotes, which literally means "before [the development of] the nucleus" (before [the development of] the nucleus), kingdom Bactéries and Archaea are often combined in a wider assembly of the prokaryotes that essentially means "before the development of" the nucleus.

Diversity and natural selection

Bacteria are different from beetles and frogs are distinct from mushrooms. It is sober, as a human being, to bear in mind that dynamic, multicellular beings are new to life on earth. Animals arrived later still and primates came late indeed. What about human development? Life form development is a process of a few millionyears in the time span of Earth's history, hardly a pin of the finger.

What is the source of its diversity when the root of a unity of life is collective ancestors? Since the variations between species are hereditary, the original cause of the differences must be a mutation. However, mutations alone do not suffice to explain why animals live in their environments, why ocean mammals have unique improvements in swimming and why desert mammals have special adaptations to make it possible to survive in small amounts of water. Mutations are happenings that are not aimed at a specific adaptive goal, such as longer fur between Arctic mammals. The process for adaptation was described by Charles Darwin (1859) in his book "The Origin of Species". Darwin contended that adaptation is a result of natural selection. Individual species with unusual changes, mutation variations that enable them to thrive in the prevailing environment or replicate more effectively, produce more offspring than other species, leading to future generations with their beneficial genes unequally. If this process is repeated over many generations, the whole species evolves genetically and a steadily growing proportion of the population inherits beneficial mutations. This includes natural selection, mutation, as well other characteristics of the genetic diversity field (Solé-Cava et al., 2013).

2.13 THE MOLECULAR GENE AND CLASSICAL GENE CONCEPTS

It has been suggested that biologists have a coherent, accurate and uniform way of conceiving genes at the molecular level, contrary to gene skepticism. The research underlying this claim starts by differentiating between two distinct ways of thinking about genes by contemporary geneticists. Classical geneticists also thought of genes as chromosome functional units, variations in which phenotypes cause differences. Today, geneticists also think of genes in this classical way, in cases where genes are defined by means of observed phenotypic variations, as the functional units in DNA whose variations cause the differences observable in phenotypes. This method of conceiving genes is considered the idea of classical genes (Waters 1994). But by invoking a molecular-level definition, modern geneticists often think about genes in a particular way. The definition of the molecular gene stems from the idea that genes are units in DNA that act in molecules synthesized through DNA expression to determine linear sequences. According to this study, contemporary geneticists are working on both principles. Two new gene definitions, which he terms "genes-P (preformationist)" and "genes-D (developmental)" are also distinguished by Moss (2003). He suggests that the convergence of these principles of genetics contributes to incorrect thought.

The fact that geneticists have often spoken about genes as if classically constructed genes are for gross phenotypic actors (phenotypes) or as if individual genes create phenotypes is due to a lot of misunderstanding about the classical way of thinking about genes. On the part of classical geneticists, this talk was very deceptive and appears to be misleading in the sense of contemporary genetics. All kinds of genetic and extra-genetic factors are involved in the development of a gross phenotypic character, such as purple eye-color, like different cellular enzymes and structures, tissue configurations, and environmental factors. Furthermore, the gross phenotypic level functions which may be assigned to each of the genes are not clear. For instance, today it is no clearer than it was in Morgan's day that the purple gene's role is to relate to the development of eye color. A number of gross phenotypic level traits are affected by mutations in this gene. Legitimate explanatory reasoning that invokes the idea of the classical gene does not rely on any baggage about what genes are for or what role a gene may have in growth. What is determined by the explanatory rationale is the concept of distinction that induces any difference in the gene, in particular genetic and environmental contexts. In contemporary biology, multiple genetic explications are better understood as regards the classical definition of genes and the theory of differentiation (Shimizu et al., 2013).

2.14 WHAT IS GENOMICS?

Genomics is the analysis of organisms' whole genomes, which combines elements from genetics. To sample, assemble, and analyze the form and composition of genomes, Genomics utilizes a mixture of recombinant DNA, DNA sequencing techniques, and bioinformatics. It varies from 'classical genetics' in that, rather than one gene or one gene component at a time, it recognizes the entire complement of inherited material in an organism. In addition, genomics concentrates on interactions within the genome among loci and alleles and certain associations like epistasis, pleiotropy and heterosis. Genomics inserts the accessibility for whole species of complete DNA sequences and was made feasible by both Fred Sanger's groundbreaking work and the more modern next-generation sequencing technologies.

In the 1970s and 1980s, Fred Sanger's group developed methods of sequencing, genome mapping, data processing, and bioinformatic analysis. In the 1990s, this work prepared the way for the human genome project, an enormous feat of global cooperation that resulted in the full human genome sequence being published in 2003. Today, sequence technologies of the next decade have led to remarkable developments in genome sequencing speed, capability and accessibility. In addition, developments in bioinformatics have made possible hundreds of databases and programs in the life sciences that provide funding for scientific research. It is possible to search, compare and review information stored and structured in these databases easily.

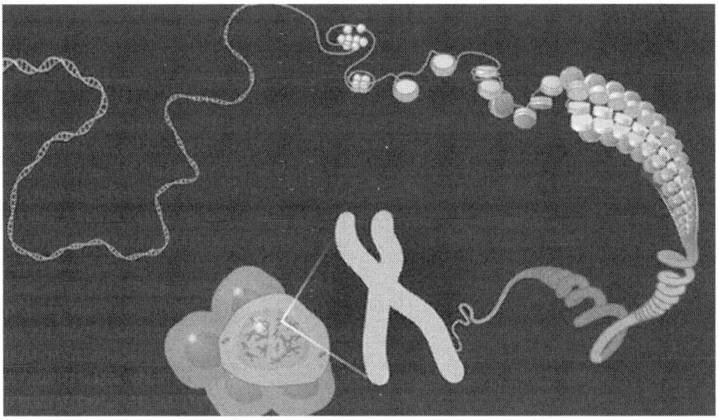

Figure 2.9: Genomics is the analysis of all the genes (genome) of an individual, and involves the interactions between them and the environment. The full collection of DNA is considered the genome of one organism. Almost every cell in the body produces a full copy of about 3 milliards of the human genome's base pairs, or letters.

Source: https://www.genome.gov/about-genomics/fact-sheets/A-Brief-Guide-to-Genomics

The study of why genes and intergenic regions of the genome relate to various biological functions is functional genomics. Usually, a researcher in this area studies genes or regions on a "genome-wide" scale (i.e. all or several genes/regions at the given time), aiming to narrow them down to a list of potential genes or regions for a more thorough study. The aim of functional genomics is to establish how a specific phenotype is formed by the individual elements of a biological system. In a particular context, functional genomics emphasizes the complex expression of gene products, for instance, at a specific stage of development or throughout a disease (Shimizu et al., 2013).

Depending on what we are focusing on, there are many particular functional genomics approaches:

- Metabolite level (metabolomics)
- Protein level (proteomics)
- RNA level (transcriptomics)
- DNA level (epigenomics and genomics)

Metabolomics, proteomics, and transcriptomics, collectively identify the biological system's transcripts, proteins and metabolites, and it is predicted that the incorporation of these data would provide a full model of the biological system under consideration. Experimentation in functional genomics tests changes in RNA (transcriptome), DNA (genome and epigenome), or DNA/RNA and protein interactions that affect a sample's phenotype. Popular functional genomics divisions include:

- Meta-analysis
- Nucleic acid-protein interactions
- Epigenetic profiling
- Transcription profiling
- Genotyping

2.15 MEDICAL SCIENCE IMPLICATIONS

Almost every human condition has a base in our genes. Until recent times, physicians could only accept biology or genetics for a small number of other disorders and congenital abnormalities. These were disorders such as anemia and sickle cells, which have very basic and predictable patterns of inheritance since each is triggered by a specific gene shift. The overwhelming amount of data from the Human Genome Project and other genome studies into human DNA gives scientists and clinicians greater power to analyze the effect that multiple biological variants play in even more complicated diseases, both in collaboration and with the environment. The majority of health issues in the United States are illnesses such as cancer, diabetes and cardiovascular. Genome-based studies have been able to enhance diagnostic, enhance treatment methods, demonstrate clinical effectiveness based on evidence and offer more efficient policy tools for providers and patients. Ultimately, therapies that are personalized to the genetic makeup of a patient seem unavoidable. The genetic basis in the health sector is thus beginning to change profoundly, with genomics being given the first examples of the period. Even so, it is crucial to note that time, effort and resources are always required to transfer results from the research laboratory to the health office. Evidently, genetics remains only one of a number of factors leading to the risk of most prevalent diseases in humans. In several cases, including several types of cancer, diet, and environmental presentation also occur (Shimizu et al., 2013).

CHAPTER 3

FISH GENETICS

CONTENTS

3.1 INTRODUCTION

Genetics is an area of biology that deals with the study of heredity and variation. Heredity relates thereto science by that living beings tally their relatives and ancestors. Variation suggests that the science by which living beings differ from their relatives and ancestors. Therefore biological science is the science to review however varied characters are transferred from parents and turn into offspring.

Of the relative multitude of organisms and plants in the marine milieu, fish is the main source of human food. They are the significant source of protein for numerous individuals and comprise the primary piece of the eating routine in numerous societies. Fish and fisheries not just give a huge part of the protein accessible for human utilization, however, they are likewise a financially critical movement, giving positions and business openings and, for some nations, a method for improving the equilibrium of worldwide exchange (Solé-Cava et al., 2013).

The inheritance of attractive qualities in yields and domesticated animals has been the establishment for specific reproducing for millennia. In any case, our understanding of hereditary qualities, which is the natural premise of heredity and variety among living beings, has developed drastically over the most recent a very long while. This new information extends the possible uses of hereditary qualities and genome advances in cultivating practices.

In aquaculture settings, current hereditary improvement programs center around choosing prevalent broodstock, utilizing better reproducing practices, expanding maintainability, and limiting ecological issues. These projects have just prompted more proficient, gainful, and productive aquaculture systems, be that as it may, the genomics insurgency vows to speed and intensify hereditary advances presently. This distribution portrays both essential hereditary standards and applied genome developments that are pertinent to the present aquaculture industry (Solé-Cava et al., 2013).

3.2 GENETIC COMPOSITION

The gene is the essential unit of inheritance. The natural code for the creation of recognizable traits, or aggregates, in a life form is contained in the gene. Genes are orchestrated on an infinite atom called deoxyribonucleic corrosive (DNA). Nucleotides are subunits of the DNA. Adenine (A), guanine (G), thymine (T), and cytosine (C) are four nitrogenous bases that characterize nucleotides. It is the combination of these units in an exceedingly linear arrangement that codes for a gene. Chromosomes are a greater part of the DNA. Chromosomes are situated in completely nucleated cells. The all outnumber of chromosomes can vary among species however will commonly be steady in nucleated cells for different species. Most amphibian species contain two arrangements of chromosomes, one acquired from the father and one acquired from the mother.

Any living being with double arrangements of chromosomes is termed diploid, or 2N. Diploids are clarified further in the article. The karyotype is that the outward presentation (size, number, state) of a life form's chromosomes. Chromosomes are of two types and exist in eukaryotic (species with nucleated cells): autosomes and sex chromosomes. Autosomes are generally chromosomes that do not seem to be sex chromosomes. Sex chromosomes generally decide the fish's gender (Vucelić-Radović, 2019).

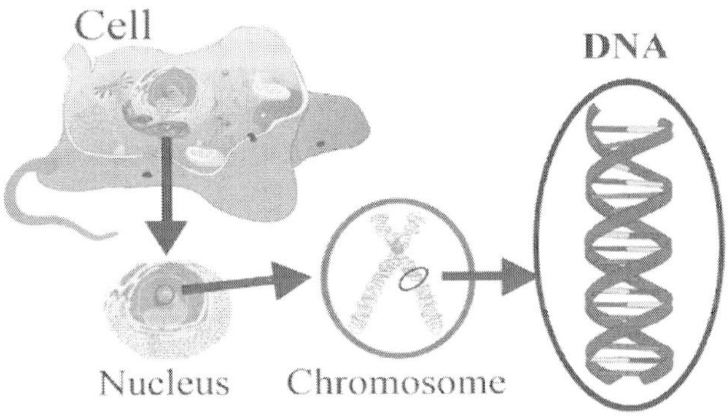

Figure 3.1. A cell and chromosome.

Source: http://aquaticnation.org/library/Lib_Gen_genetics02.php

A number is typically assigned to an autosome, while the trademark documentation for sex chromosomes are XX and XY in diploids, in which case the XX are females and XY are males within the XY sex-assurance framework. Numerous kinds of fish can have an alternate sex-deciding organization. Autosomes may be solely contained in some while other fish contain all the sex es and either change sex as they develop or, in uncommon cases, have male and female gametes at the identical time. Resolution of sex chromosomes is absent in certain fish species.

In fish species, there are exemptions for the quantity of chromosome sets (ploidy), although most eukaryotic cells are diploid. In hereditary improvement programs, the quantity of chromosome sets can be subject to control. The creation of haploids (N) is possibly similar to fish possessing chromosomes from the mother (gynogens) or from the father (androgens). There are three arrangements of chromosomes (3N) referred to as triploids, and four arrangements of chromosomes (4N) are referred to as tetraploids. Naturally, tetraploids tend to be more regular than triploids. An example of a diploid (2N) species is the channel catfish, it has 29 sets of chromosomes (2N = 58). Salmon and trout, the salmonids, and the conventional carp are broadly seen as tetraploid (4N) (Vucelić-Radović, 2019).

Due to the fact that chromosomes exist two by two in diploid organisms, every quality has, in any event, two duplicates. Each duplicate of quality is referred to as an allele, and it is situated at a particular area on every one of the sister chromosomes. This area is assumed because of the locus of the allele. Two alleles could also be indistinguishable or can have varieties in their DNA arrangement. If alleles at a particular locus are indistinguishable, the living being is homozygous for that quality. On the alternative hand, if alleles at a particular locus are unique, at that point, the organism is heterozygous for that quality. Alterations in alleles can create hereditary changes and possible evolution in a population. The mixture of alleles for a specific trait is known because of the genotype (Vucelić-Radović, 2019).

3.3 CELL DIVISION AND THE REPLICATION OF CHROMOSOMES

Mitosis and meiosis are the two different forms of cell division. During the development of life forms the separation of cells occurs and this is repetitive leading to an increment in number, supplanting the old and dead cells. As a life form develops, cells should separate and repeat to increment in number and to supplant old or dying cells. In the Mitosis cycle, substantial cell partitions deliver two indistinguishable little female cells. Large cells aside from egg and sperm and the cells that produce them come from physical cells. When mitosis occurs, the whole cell is copied, and this includes the chromosomes and any remaining cell mater. Along these lines, every female cell consists of a diploid arrangement of chromosomes.

The mitosis cycle contrasts with meiosis, particularly the gamete (egg and sperm) arrangement. Cell division occurs in the case of both the eggs and the sperm and their replication results in cells containing a haploid (N) gamete, possessing just one bunch of every pair of chromosomes. The way toward making haploid gametes is basic to multiplication in diploid living beings. The right quantity of chromosomes (2N) is recuperated in the zygote at the point where fertilization of the egg by the sperm. The process of producing a sperm cell by gametes is referred to as gametogenesis while the production of egg cells is referred to as oogenesis (Caliskan et al., 2015).

At any rate, in meiosis, three significant cycles happen to create hereditary inconstancy in the sperm and egg: getting over, isolation, and free grouping. To start with, the chromosomes in the gametocyte should go through the process of replication. After diploids are recreated from chromosomes, the homologous chromosomes structure sets in packs of four, called quadruplicates. Now, the chromosomes that form quadruplicates can fold over one another. The chromosomes may then split up and pieces from various chromosomes therefore rejoin. Subsequently, hereditary components have been exchanged starting with one chromosome then onto the next.

A bit of a chromosome from the mother is exhanged between its homologous chromosome from the father and the other way around. Homologous chromosomes cross over in the process in the cycle through which the actual exchange takes place. The process is a cell stage referred to as essential gametocyte occurs (essential spermatocyte in males, essential oocyte in females).

Different cycles that boost hereditary variety incorporates isolation and free collection. The chromosome complement must be decreased from the diploid to the haploid state after the chromosome is duplicated and reaches the gametocyte during the process of cell division where both the egg, and sperm are included.

Every auxiliary gametocyte comprises of a solitary duplicated chromosome coming from an individual homologous pair. For oogenesis, one of the chromosomes of each pair goes to an auxiliary spermatocyte while in spermatogenesis, all chromosome sets are discrete. Each pair of chromosomes goes to an additional oocyte or a polar body in addition to all chromosome sets being seggregated. The reason for Mendel's law of seggregation is based on the cycle in which the copied chromosomes are seggregated and form gametocytes (Caliskan et al., 2015).

Mendel's law of independent assortment states that each gamete gets an indepedant combination of alleles, for every locus, coming from the parents. The seggregation cycles and independent assortment are irregular that means that each auxiliary gametocyte or polar body comes from the mother and the father and each gets a random combination of chromosomes.

A reduction of chromosomes from the essential gametocyte to the optional gametocyte diminishes the quantity of chromosomes from the diploid complement to the haploid complement. This cycle guarantees that upon a sperm fertilizing an egg, the right complement of chromosomes are created in the zygote (prepared egg). The seggregation for the autonomous variety of chromosomes (and alleles) is linked. On the off chance that the qualities that control a trait are situated on a similar chromosome and are near one another, they are supposed to be truly connected. For this situation, these qualities might be acquired at the same time.

The auxiliary gametocytes comprises of two optional spermatocytes or an auxiliary oocyte and a polar body. Every auxiliary oocyte ordinarily delivers just a single haploid egg; the remaining are polar bodies. In females, following isolation and free variety, chromosomes toward a single side are squeezed off with a touch of encompassing cytoplasm (yolk). It results in structuring of the main polar body. The non-useful Polar body cells are a result of oogenesis. Given the egg is the significant site of sustenance to the undeveloped organism, a high centralization of cytoplasm is fundamental and henceforth the inconsistent separation of cytoplasm in the oocyte compared to the polar body (Bert, 2007).

The principal polar body could conceivably isolate once more to deliver two little haploid cells. The other cell is the optional oocyte. The auxiliary oocyte and first polar body produce one developed egg and one optional polar body. Optional oocytes are not stimulated until they have reached the sperm. For males, the two auxiliary spermatocytes result in the development of four sperm cells. The sperm cells possess a haploid number of chromosomes that are autonomously different and have equivalent measures of cytoplasm.

3.4 THE DEVELOPMENT

After the process of fertilization, the delivery of a diploid zygote occurs. The outcome is an individual cell; mitosis of the zygote and the ensuing little young female cell is answerable for the development and development of the fish. A yolky cytoplasm is mostly found in fish eggs. As a rule, following preparation cell division starts without the blastodisc, which is the yolk are of the egg. The underlying cell division is referred to as the cleavage stage and the underlying cells are named blastomeres. A blastoderm is formed from the division and expansion of blastomeres.

The initial cells blend and can offer ascent to an assortment of cell and tissue varieties. After a time of quick cell cleavage, the slow process of cell division commences, and cell development starts. Blastomeres start to group, and isolate all through the undeveloped organism. After the blastoderm has filled about a large portion of the yolk or sooner relying upon the species, germ layers start to frame through the cycle of gastrulation or cell rebuilding. These layers will form into the tissue and organ systems of the fish. From a hereditary angle, zygotic trait expression by and large starts during the mobile division stages (Bert, 2007).

3.5 SEX DETERMINATION

Sex qualities are the principal determinants of sexual orientation. The sex chromosomes of various fish species might be morphologically unmistakable. Many sex chromosomes take after autosomes, as well as some sex-deciding qualities might be situated on autosomes. In these cases, sex-connected traits should be concentrated to decide the sex chromosomes or then again sex qualities that are situated on autosomes. The XY diploid framework is the most widely accepted sex determination framework.

This framework is well-known in referring to fish species just as it does in relation to humans. The XY framework can be observed in Channel catfish. This framework was so named because the sex chromosomes in people take after a Y or an X. for the framework XY, the sex chromosomes in females are indistinguishable (XX), while those in males are a blend (XY). The Y chromosome is discovered distinctly in males. This is because the sets in females are the equivalent, these chromosomes are named homogametic, and are referred to as heterogametic in males (Lee & Donaldson, 2012).

A solitary X chromosome is contained in all eggs. For the male, a large portion of the sperm populace contains a solitary X chromosome and half contain a solitary Y chromosome. In the XY framework, the heterogametic sex (male) is the one answerable for sex determination in the population. On the off chance that an egg (X) is fused with a sperm carrying the X chromosome, the gender of the offspring will be female (XX). On the off chance that an egg (X) is treated with a sperm carrying the Y chromosome, the offspring will be male (XY).

Once more, the consequence of offspring is diploidic. Likely, just a bit of the sex chromosome is at last liable for sex assurance. Different traits (qualities) might be situated on the XY chromosomes in various aquaculture species. There are a few other sex-deciding systems in fish other than the XY framework. One is the WZ framework. It functions precisely the inverse to the previous framework; a homogametic fish is a male whereas a heterogametic fish is a female. The WZ sex determining system can be observed in Blue tilapia. A different set of chromosomes found in the female blue tilapia is (WZ), contrary to this the males have the (ZZ) sex chromosomes. The chromosome that decides the sex is the W chromosome the sperm that fertilizes the egg will then create the male progeny (ZZ), and if the sperm fertilizes an egg with the chromosome type (W) then a female (WZ) is produced. Given this, the gender of the fish is governed by various egg types and just as the XY sex determination system (Lee &Donaldson, 2012).

ZO is another existent system for sex determination. Examples of fish with the ZO framework include the midget gourami fish, and a sole fish. The females are ZO, and the males are ZZ. The sex deciding species are the females and are the heterogametic sex. By any means, haploid eggs are produced by the female with no sex chromosome or a Z chromosome.

If a sperm (Z) fertilizes an egg with the Z chromosome (Z), at that point, the offspring is the male (ZZ). A female offspring (ZO) is delivered in the event the egg has no sex chromosome (O). This XO system likewise is existent in certain fish varieties, counting sunfish animal varieties. For such a system, the males are heterogametic (XO) deciding the sex of the offspring.

As it were, the XO sex-assurance framework is like the XY sex-determination system aside from the one existent XO system only a set of sex chromosomes is contained in the males.

There exist other sex-deciding systems in fish. A large number of the systems are more intricate than the XY framework and its variations. The other systems have different sex chromosomes.

An illustration of a multiplex framework is the WXY sex-deciding framework, conventional in the platy fish utilized in aquaria. In this case, the two males and females can be either homogametic or heterogametic deciding the gender of the fish. The W chromosome goes about as an adjusting chromosome and is able to hinder the male-deciding capacity of the Y chromosome in WXY. Hence, the XY and YY offspring are males, and the XX, WX, and WY offspring are females (Lee &Donaldson, 2012).

Sex deciding systems can also incorporate the X1X1X2X2/X1X2Y, ZZ/ZW1W2, and XY1Y2/XX systems. At long last, sex-deciding qualities additionally can be situated on autosomes. Some aquaculture species don't have any morphologically unmistakable sex chromosomes. In those species, sex is controlled by a few females or male qualities situated on specific autosomes.

One significant idea to present is sex based on phenotype versus heredity. Since Fish are lower vertebrates, they possess have a range of adaptability in sex organ improvement, so phenotypic sex might not properly represent chromosome formation of the fish. A few elements can impact the declaration of sex aggregate. The absolute basic elements are temperature, saltiness, population density in tanks, pressure, hormone therapies, radiation, and photoperiods.

Sex changes might be initiated or are likely to happen normally in certain populations. For example, in the use of estrogen for hormone treatment often creating males in some fish species can deliver phenotypic grown-up females despite the fish contains the XY chromosomes. Then again, hormonal therapy can prompt the development of most of the males despite the fish having female hereditary traits (XX chromosomes). A good example is the typical practice in aquaculture is the treatment of tilapia with 17α-methyltestosterone (MT) to activate each single male population. Female tilapia develop slower than their male counterparts, such hormonal therapy prompts expanded yields. In such cases, nonetheless, the fish may harbor XY chromosomes (hereditary males and phenotypic males also), they might also have XX chromosomes (hereditary females, yet phenotypic males) (Beaumont &Hoare, 2008).

3.6 QUALITATIVE GENETIC TRAITS

Phenotypes exist under the class of subjective qualities or characteristics basically portrayed as either male or female. A characteristic is subjective if it very well may be arranged into one of at any rate two classifications. These traits are not estimated over a range, just like the case with numeral characteristics. An illustration of a subjective characteristic would be the pigment on fish; the fish is either typically pigmented or a pale-skinned individual. Subjective qualities are regularly the least difficult to assay because they are probably going to be constrained by just one or a couple of qualities, in contrast to quantitative traittraits.

Since a aquaculture species can be governed by these traittraits, the population can be portrayed by the proportions of its individuals with these qualities. The quantity of fish with every traittrait is essentially added up and portrayed, for example, 3:1, 1:1, etc. For instance, a populace of 20 fish has either a fan tail or a round tail, a subjective characteristic. In the event that 15 fish have fan tails and five fish have round tails, the proportion of fan tails to adjust tails is 3:1. Portrayed another way, one-fourth of the fish population has round tails and seventy-five percent of the population have fan tails. Subjective traits are equivalent to various conditions (Shimizu et al., 2013).

Various components impact these proportions, including the number of qualities expected to deliver the aggregate and quality activity. The hereditary factors portrayed here are normally alluded to as old-style hereditary qualities or Mendelian hereditary qualities. The quality activity can be described as single or various qualities creating a subjective aggregate or phenotype. Classic subjective traits are predominant or latent. On the off chance that a solitary allele is carried over over the other at a similar locus, at that point the method of activity is named predominance.

The complete strength method of activity portrays the statement of alleles at a similar locus in which case a duplicate (the dominant allele) covers the impact of the later duplicate (the recessive allele). The trait carried over is the dominant trait and the latter is the recessive trait. At the point when dominant alleles happen, three genotypes are conceivable, while just two combinations can be delivered. An exemplary illustration of the comprehensive strength of a solitary locus is the quality of albinism in the channel catfish.

The dominant genotype resulting in ordinary catfish with pigments can be named AA or Aa, where (A) is the dominant and (a) is the recessive. Both AA and Aa fish produce ordinary pigmentation. Parents with the total latent color quality (aa) will be a pale-skinned individual. Thus, there are three blends of genotypes (AA, Aa, aa), while just two aggregates are delivered—ordinary (AA or then again Aa) or pale-skinned individual (aa) tinge. Homozygous fish dominant (AA) will be regularly pigmented, however so will fish with one dominant (A) and one passive (a) allele (heterozygous dominant, Aa). With complete strength, parents that convey recessive alleles (either Aa or aa) can create offspring with a latent characteristic (Shimizu et al., 2013).

There are two different prospects concerning subjective qualities—various alleles and sex-connected alleles. At the point when alleles at more than one locus control a subjective aggregate, there are two prospects of quality activity. There are either epistatic impacts or non-epistatic impacts. Epistatic impacts are cooperation between qualities that can cause adjustments or concealments of aggregates. Accordingly, mixes of numerous alleles at various loci can cause unexpected combinations in comparison to the basic instance of either/or examples.

3.7 PHENOTYPIC EXPRESSION COMPONENTS

In a subject, a quantitative characteristic is resolved by its qualities, the environment, and the communications of the two. This relationship is:

$$P = G + E + GE$$

Where: P is the phenotype for a communicated total G is the impact of genotype; E is the value for environmental impacts; GE is the joined impact of the environment and genetics on the person's genotype. Hereditary qualities and environment associate contrastingly in each individual. An individual might be affected by contrast in various conditions. Various individuals may react contrastingly to a similar environment. Critically, the individual genetic value can be separated into its central segments. The genetic value (G) can be depicted as the amount of additive impacts (A), dominance effects (D), and epistatic effects (I) with the end goal that:

$G = A + D + I$ Additive impacts are because of the total impacts of alleles at all loci for a given trait. This worth is free of different associations among alleles or different blends of alleles, so additive impacts are protected via meiosis Furthermore, these qualities are typically passed from parents to offspring. Predominance impacts are because of the cooperation of alleles at every locus. In the process of meiosis, homologous chromosomes are isolated and decreased from the diploid to the haploid status. Given the gametes are in a haploid state, they can possess no domnance impact; new matches of alleles structure after treatment to make new and irregular diploid states. Accordingly, predominance impacts are made in various mixes with every age. Epistatic impacts are additionally significant in hereditary condition. These impacts are brought about by the collaboration of alleles at various loci. Expressed, this is the impact of cooperation between qualities. During meiosis, epistatic impacts for most qualities are upset by seggregation and independance (Dunham, 2011).

Thus, most epistatic impacts are reproduced with every age. Since quantitative qualities have dissemination and difference in a populace, one approach to comprehend these characteristics is to examine their change and impacts. Reproducing programs can damage the hereditary change in a population all together to limit its impact. The condition for phenotypic impacts can be adjusted from the individual to be pertinent to the population.

The best approach to assay quantitative qualities in a population is to read the difference for an trait or characteristics. Change is presented from the environment or environmental variance (VE), hereditary qualities or genetic variance (VG), and the collaboration among hereditary qualities and the environment (VGE). Hence, the total phenotypic variance (VP) for any quantitative quality is:

VP = VG + VE + VGE

The hereditary difference is utilized in reproducing projects to aid the choice of a supply of fish. To consider genetic variation, the value should be separated into its chief parts.

VG = VA + VD + VI

here VA = additive variance (variance from additive gene effects), VD = dominance variance (variance from dominance gene action), and VI = interaction (variance from interaction between genes)

The connections and contrasts among this variance value, how they are acquired, and their corresponding sums are significant in any rearing program (Dunham, 2011).

3.8 GENETIC VARIANCE AND BREEDING

Hereditary difference is a vital part of misusing the characteristics of interest in a reproducing program. From the condition above, segments of hereditary change incorporate epistatic impacts, additive impacts, and predominance impacts. Epistatic impacts are infrequently útilized in a rearing project since there are countless potential mixes of alleles in this estimation of variance. Most projects zero in on additive and strength hereditary change. The relative commitment of both of these impacts on the phenotype of interest decides the best way to deal with use in a rearing system.

Additive impacts are transferred from parent to offspring and predominance impacts are made with each age. If additive impacts are huge, and accordingly the variation inside a populace is enormous, fish with an alluring characteristic can be chosen and reproduced. In the event that the additive hereditary difference is little, at that point choosing fish for a particular characteristic inside a population may not deliver offspring that express the character better than the parents (Beaumont et al., 2010).

For this situation, in the breeding program, the alleles are recombined through the conventional would be the normal method. Additional alleles are presented through a flood of new hereditary material, in light of experimentation. New strains of fish can be brought into the reproducing program (intraspecific hybridization) or various species reproduced (interspecific hybridization). Another mix of alleles may create offspring that harbor the ideal mix of phenotypes.

For a determination program to be successful, the measure of additive change should be resolved. The proportion of heritability (h2) portrays the extent that additive hereditary change adds to an aggregate in a population. Heritability portrays the level of an aggregate that can be acquired in an anticipated way. This value should be dependable since, for a quantitative aggregate, the genotype isn't influenced by meiosis (Beaumont et al., 2010).

These qualities range from 0 to 1 and are a rate. Heritability estimation of 1 recommends that the aggregate noticed is 100% clarified by the additive variance. Heritability estimation of 0 recommends that additive hereditary impacts don't add to the average amount. All in all, heritability esteem more noteworthy than 0.2 (h2= 0.2) proposes that a trait might be dependably misused in a determination program. If a characteristic's heritability is lower than h2= 0.15, a choice-based program may not be the appropriate response since strength hereditary difference is more significant for this situation. Simply recall that we are managing quantitative characteristics here. These characteristics might be influenced by the climate and the populace and can differ between ages. Direct choice of traits with low inheritance might be troublesome; in such circumstances, genome-based innovations, for example, the marker-helped determination might be more impactful (Beaumont et al., 2010).

3.9 GENETIC IMPROVEMENT PROGRAMS

a) Selection:- The hereditary difference is a vital part of misusing the characteristics of interest in a reproducing program. From the condition above, segments of hereditary change incorporate epistatic impacts, additive impacts, and predominance impacts. Epistatic impacts are infrequently utilized in a rearing project since there are countless potential mixes of alleles in this estimation of variance. Most projects zero in on additive and strength hereditary change. The relative commitment of both of these impacts on the phenotype of interest decides the best way to deal with use in a rearing system (Solé-Cava et al., 2013).

Additive impacts are passed down from parent to offspring and predominance impacts are made with each age. If additive impacts are huge, and accordingly the variation inside a populace is enormous, fish with an alluring characteristic can be chosen and reproduced. In the event that the additive hereditary difference is little, at that point choosing fish for a particular characteristic inside a populace may not deliver offspring who have a better expression of the character compared to their parents. For this situation, in the breeding program is through conventional methods. More alleles are presented through a flood of novel hereditary material, in light of experimentation. New strains of fish can be brought into the reproducing program (intraspecific hybridization) or various species reproduced (interspecific hybridization). Another mix of alleles may create offspring containing the ideal mix of phenotypes.

For a determination program to be successful, the measure of additive change should be resolved. The proportion of heritability (h2) portrays the extent that additive variance adds to an aggregate in a populace. Heritability portrays the level of an aggregate that can be acquired in an anticipated way (Solé-Cava et al., 2013).

This worth should be dependable since, for a numeric aggregate, the genotype isn't influenced by meiosis. These qualities range from 0 to 1 and are a rate. Heritability estimation of 1 recommends that the aggregate noticed is 100% clarified by the additive variance. Heritability estimation of 0 recommends that additive variances don't add to the aggregate.

All in all, heritability esteem more noteworthy than 0.2 (h2= 0.2) proposes that an trait might be dependably misused in a determination program. If a characteristic's heritability is lower than h2= 0.15, a choice based program may not be the appropriate response since strength hereditary difference is more significant for this situation. Simply recall that we are managing quantitative characteristics here. These characteristics might be influenced by the environment and the population and can differ between ages. Direct choice of traits with low heredity might be troublesome; in such circumstances, genome-based innovations, for example, the marker-helped determination might be more material (Solé-Cava et al., 2013).

b) Hybridization:- With low heritability, selection strategies become the most ideal approach to expand performance. Improving desirability in this context requires a more noteworthy hereditary variety for novel mixes of alleles is required. Novel mixes can be made through fish reproduction with various hereditary narratives, a cycle called hybridization. All offspring is called half and halves. The parents can be of similar species, yet various strains (intraspecific hybridization). Or on the other hand, the parents can be of various species (interspecific hybridization).

Improving a characteristic utilizing hybridization is undertaken via experimentation. A portion of the half and halves may have prevalent characteristics and while others lack them yet hybridization seems to be the most ideal way to expand execution when the heritability for a characteristic is exceptionally low since hybridization makes new allele combinations through dominant variance. This measure is autonomous of heritability, ensuring hybridization projects can be utilized in any event, if there is high heritability. Selection and hybridization can both be utilized to increment fish performance.

I. **Intraspecific hybridization:-** This form of hybridization can be utilized to enhance how fish perform in different ways: It very well may be utilized to develop another strain that can go through choice for an trait, and it can likewise be end result with the mixtures being the finished result. All in all, mixtures have better wellness given more prominent hereditary changeability. This is known as half breed power or improvement through outbreeding.

To make another variety, a cross should be made between people of two strains with various hereditary backgrounds, trailed by a defining program to better its execution. A choice program utilizing cross breeds typically can't start until the second generation of fish (F2) is produced. This is expected to a great extent to the rule of strength hereditary variety; original mixtures (F1) by and large try not to pass on the mixture energy (and predominant traits) to all offspring.

The half breeds are made utilizing two particular strains and, thus, additive hereditary impacts should be expanded. This implies that the resulting ages are reasonable for utilization in the program through selection. Another methodology is to direct a choice program inside a strain and afterward use hybridization to attempt to improve execution. Nonetheless, not all crossbreeds will express enhanced execution. Achievement relies not just on the strains chosen yet also on the equal hybridization. This implies that a male of one strain crossed with a female of an alternate strain may create descendants with various characteristics, and the other way around. The crossbreeds that play out the best will be dictated by experimentation.

II **Interspecific hybridization:-** Various species of fish might be crossed to deliver more gainful descendants. Interspecific hybridization is normally utilized in the misuse of the mixture power, or the propensity of a crossbred organism to have characteristics better than those of one or the other parent. Nonetheless, if descendants are prolific, interspecific hybridization has likewise been utilized to better hereditary qualities using introgression, a cycle by which the qualities of one animal type stream into the genetic stock of another. This is accomplished by backcrossing an interspecific half and a half with one of its parents.

Similar standards apply likewise with intraspecific hybridization: Hereditary improvement depends on new mixes of alleles. Interspecific mixtures should have the option to deliver descendants. Whenever this is set up, a rearing system can endeavor. Yet, numerous between-species mixtures are sterile, try not to recreate as promptly as the parents, or produce descendants that are non-reasonable or irregular. Spawningdoes not happen normally between numerous species that can be hybridized, so these species should be misleadingly produced. The way toward delivering a predominant fish includes exploring different avenues regarding mixes of strains, species, and equal reciprocal crosses.

A proper illustration of fruitful interspecific hybridization is the cross between the channel catfish and blue catfish. Every one of the species type has a few distinctive unrivaled qualities. Channel catfish is best for business creation on account of its development rate. The Blue catfish has a more uniformity in its appearance, yields more filet, is simpler to seine, and is more impervious to specific infections. Genetic crossing between female channel catfish and male blue catfish results in suitable offspring. Thehaploids have a short development time, and are safer from illness compared to the parent species. With this information, hybridization projects can deliver fish subject to be chosen for additional hereditary upgrades

c) Polyploidization:- The majority of fish species are diploid and have a pair of chromosomes (2N), one set derived from the mother and the other set acquired from the father. The diploid number of chromosomes is less than those of polyploids. Polyploidy is actuated in fish by utilizing strategies, for example, temperature variety or application of weight to the eggs to make triploids (3N) or tetraploids (4N). Shocking recently prepared eggs result in the development of triploids. When shocked, the egg doesn't get rid of the subsequent polar body.

The result is a prepared egg from the sperm and the others from the egg and the polar body. During development, the haploid nuclei which are three will mix and make a triploid. Heat shocks can also induce this. Triploids are made to expand fish development and to govern populations by initiating sterility. Triploids ought to become bigger as the cell size increases (containing more hereditary material and bigger nuclei). Furthermore, given triploids are typically clean, a lower effort is expected to create gametes, what's more, this energy might be redirected to development. The ordinary number of chromosomes is disturbed and with an upset independant assortment making the triploids sterile. This complicates the process of gametogenesis since the separation of the chromosomes is not possible (Solé-Cava et al., 2013).

In fish cultivation, triploid grass carp are frequently developed in lakes alongside different species to keep grass and weed under wraps. While there are restricted uses for triploids in large scale aquaculture, utilizing the cycle for species with very high fruitfulness, for example, shellfish, has indicated a great degree of achievement. Ranchers need huge quantities of eggs since numerous eggs don't endure taking care of and shock. Another approach to deliver triploids is to mate a diploid with a tetraploid that has been created.

This strategy is likely to increase reasonability, yet it needs a tetraploid population. Tetraploids have four sets of chromosomes (4N). Their creation is through the shocking a zygote when it is going through mitosis. After duplication, the shock can or should be applied, and as the nucleus is separated into two. Four chromosomes are held given the shock keeps the core and cell from dividing.

The single motivation for delivering tetraploids is to make triploids, as referenced previously. Triploids are made more productively by mating diploids with tetraploids than they can with shock treatment, as this causes critical damage. Numerous tetraploids can deliver offspring, so once a populace of tetraploids is created, they could proliferate without making another population without fail (Rowe et al., 2017).

d) Production of gynogens and androgens:- Gynogenesis is the formation of haploid fish with their DNA from their female parent. This ubiquitously entails the fusion of sperm subjected to UV irradiation with the oocyte to stimulate division in the latter. Though the DNA of the sperm is destroyed, it can enter the egg to induce division with polar body formation. This generates haploid organisms.

The restoration of diploidy involves pressure shock or low or high temperature treatments. The possibility of recombination among the chromosomes of the ovum makes these meiotic gynogens not entirely homozygous despite the exclusively maternal origin. The use of chemicals or pressure or temperature prevents the separation of the second polar body to thus avoid diploidy. In the case of mitotic gynogens, the DNA is exclusively 100% from the mother. This is of value for lowering variances and creating an all-female population in the case of the XY system.

In the case of angrogenesis the source of DNA is the father with UV treatment of the egg. A haploid zygote is the outcome from this treatment. The diploidy is restored by temperature or pressure shocks at the first instance to multiply the paternal haploid set. As they are exclusively homozygous, they are multiplied haploids. They find use in an all-male (XY) population. This population in tilapia sometimes demonstrates higher rates of development (Rowe et al., 2017).

e) Genetic engineering :- Genetic engineering is the cycle by which a gene(s) or then again a useful piece of quality is moved into a living being. The quality are likely to originate from similar species as the beneficiary or alternate animal varieties. A transgenic fish is delivered upon fruitful quality exchange. The desired trait is then proliferated in offspring. A few cycles are engaged with genetic engineering. Initially, the trait we are interested in is cloned and embedded into a vector, for example, a bacterial plasmid DNA. The amounts of plasmid from the microbes are huge.

The quality of interest, or the DNA embedded into the plasmid, is eliminated from the vector and infused into the fish zygote, where it is normal that the novel quality will turn out to be important for the host fish DNA. When the moved DNA is joined into the germ cells, it is acquired as a piece of the genome; fish so created are transgenic fish. Genetic engineering has the benefit of altering the boundary of the species.

It maintains a strategic distance from epistasis by moving a particular gene(s). Hereditary control components additionally can be controlled to permit a quality to be constrained by an alternate promoter, as wanted. Such advertisers can be induced. For example, a development hormone quality can be put heavily influenced by a constitutive promoter, prompting the constitutive synthesis of high development hormone, which instigates quick development.

An inducible promoter can be utilized to recognize specific impurities for ecological observing. For example, P450 oxidase promoters are inducible upon presentation to specific chemicals. Transgenic fish with these promoters, together with a marker, would communicate the marker when the transgenic fish is exposed to the chemicals. Despite these favorable circumstances, genetic engineering is a unique methodology and the vulnerability of its impact on both safety and natural security has created much fuss from its application in aquaculture. Until this point in time, transgenic fish have had restricted use and restricted monetary effect. The all-around recorded commercial utilization of transgenic fish is the decorative gleaming zebrafish, the GloFishTM, made in Singapore and industrially accessible in the US.

Even though the innovation for creating transgenic fish is developed, the creation and confirmation measure is lengthy, and the expenses are still high. Significant snags for utilizing transgenic fish are social obstruction and guidelines. Subsequently, ongoing investigations have zeroed in on surveying transgenic fish as far as food and environment security, and on approaches to contain transgenic populations.

f) Molecular genetics and genomics:- Molecular genetics is an arising scope in fish rearing projects. This includes the investigation of hereditary components (genotypes) to assist in deciding whether fish have certain traits of revenue (aggregates). A means for hereditary testing that will before long become practical for the aquaculturist is DNA marker-helped choice. At the point when a specific interesting characteristic is to be considered, and a hereditary marker for this trait, a DNA test can figure out which fish in the population that can be based used for reproduction.

Some rural projects, for example, meat and poultry have executed these developments in their choice and reproducing programs as of now. As our insight into the hereditary qualities of aquaculture species, genetic experiments will turn into a reality for this industry too. The genome is the whole DNA arrangement of a life form.

Genomics is the investigation of the whole genome or DNA of an animal variety and how qualities collaborate inside the entire life form. Entire genome successions for some significant aquaculture species should be known without further ado as a result of advances in sequencing innovations, however, the total DNA succession of most aquaculture species is by and by obscure.

Genomic projects may at present be helpful for any rearing system. Guides of helpful traits (their situation along the chromosomes) will be important in the coordination of genomic information with conventional determination programs. The drawn-out objectives of a genomics program are to recognize sets of traits and have the option to plan different creation characteristics to their chromosomes to aid determination.

A comprehension of the standards of hereditary qualities is valuable in any aquaculture program. Genetic programs can assist increment efficiency in aquaculture systems, as in utilizing hybridization and choice to deliver strains with predominant desirable traits. Hereditary qualities research has advanced the quality and creation of marine species consistently, yet there is a requirement for additional and quicker hereditary increases.

A significant part of the development to date has been made utilizing specific determination. As submolecular hereditary qualities also, genomics instruments and innovations are executed in aquaculture systems, they will supplement and expand traditional genetic improvement programs (Overturf, 2009).

CHAPTER 4

GENETICS AND FISH BREEDING

CONTENTS

4.1 INTRODUCTION

While genetic experiments have been undertaken for genetically related fish species, such as salmon, for several thousand years, there has been very little genetic tinkering of any fish species in formal aquaculture facilities until the end of the 20th century. As a result, the bulk of aquatic and farm animals are very similar to wild organisms. Genetic enhancement programs are beginning to expand to a rising number of marine species, but the aquatic sector remains far behind the domestication rate of livestock and crops. The Food and Agriculture Organization of the United Nations for the Compilation of Mediterranean Aquaculture Data had just begun sea bream and marine bass farming in 1984, with just six species primarily shellfish and a few mules. There are currently 20 to 30 Mediterranean farmed species for which data are reported; some of the main species grown in the region are high-value marine species. Genetic alteration approaches are used if more and more species husbandry advances. As the demand for aquaculture increases, prices for many aquaculture products have decreased. However, there is a need for increased efficiency of production and diversification of aquaculture products to ensure that the sector remains economically competitive. Genetic science is one of the methods of maximization and diversification. Expansion of the aquaculture industry has given rise to concerns about the environmental threat and natural habitats by large-scale release or escape and subsequent proliferation, competition or predation of local ecosystems; farmed aquatic animals may also be a source of disease transmission to wild stocks (Solé-Cava et al., 2013).

Figure 4.1: Fish genetics

Source: https://diszhal.info/english/articles/fish_genetics.php

Aquaculture is the main source of purposeful migration (introduction) of marine organisms, and experience has demonstrated that cultivated plants prefer wildlife migration. The inclusion of maladapted genes, i.e. farm-specific genes and not local, hatchery or farmed fish, is believed to pose a threat to the natural gene pool. However, it was found in 1995 that the convergence of hatchery and ecology made it difficult to obtain strong proof of detrimental effects. Environmental biosecurity will be more concerned as genetic techniques broaden their use and change their phenotypes. The international community has recognized the importance of establishing the genetic principles of aquaculture in drawing up the FAO Code of Conduct for Responsible Fisheries. The following general technical recommendations have been drawn up to make this paper more successful, covering the selection and treatment of breeders, the release of GMO, the interaction between wild and farmed animals and the use of species added. Article 9.3 points out that States must, by proper management, protect and maintain the dignity of marine communities and environments. The goal of the paper is to draw attention to some representative examples of genetic technology used in aquaculture and to highlight some of the major issues related to sustainable aquaculture production (Solé-Cava et al., 2013).

4.2 GENETIC IMPROVEMENTS

Genetic advances cannot be used for a multitude of reasons except to improve yield in aquaculture. Appropriate genetic technologies may promote improved marketability, management over cultural and natural resources. Genetic engineering programs can be used with short-or long-term gains. Short-term effects are mostly immediate, over two decades and often not progressive (unless combined with other long-term measures), whereas long-term programs, such as selective breeding, provide minimal advantages for each generation. Furthermore, the evolution of genetic improvement has not been used for aquatic animals at the same level as for freshwater and anadromous salmonide organisms. Genetic developments, with more aquatic species appearing, can also be expected to play an increasingly important role in the production of productivity.

4.3 LONG TERM STRATEGIES

There is no doubt that only long-lasting breeding schemes using additive genetic diversity can achieve domestication and maximum capacity for the use of marine genetic resources. The Norwegian Atlantic Salmon Breeding

Program is an example of a long-term breeding strategy that has been reflected in other regions of the world, such as increasing tilapia in the Philippines and producing coho salmon in Chile (FAO, unpublished report). Selective breeding in Japan has been in service for the improved production of red sea bream for more than 30 years (7 generations) (7 generations). The time required to produce market-sized fish was shorter (1 kg) (1 kg). The characteristics considered for production should be of economic significance in the aquaculture market.

While the growth rate is more commonly modified in selective breeding systems, it has been observed that other characteristics are more likely to be genetically additive and increase. Disease tolerance is an essential characteristic in some aquaculture species, but it is difficult to improve this trait by unique selective breeding schemes and not through additive variety problems due to test-design problems (such as fish that are not disease-resistant or have not already undergone disease?).

In recent years, the industry has seen other characteristics, such as skin durability, due to the high cost of high-quality products on the market. When the proportion of additive genetic variation associated with locus or loci exceeds the legacy of a function, specific gene loci, known as a marker, will optimize the selection efficiency. However, it is also difficult to identify beneficial loci in aquatic organisms (Conceicao & Tandler, 2018).

Micro-satellite probe is widely genetically modified and can contribute to identification and pedigree in wide areas in mixed family groups. However, for commercial and small-scale fisheries, marker-assisted selection could be necessary because the everyday process of fishing cannot be disrupted or because there is no need for special reproductive facilities and there is no need to mark fish.

Limited tissue sections and fine clips that may not compromise marketability can be genetically studied. In general, the negative effects of inbreeding should be taken into account in the planning and execution of selective aquaculture breeding projects and the creation of family relationships with genetic markers such as microsatellites, etc.

Transgenic fish production is a long-term reproductive strategy when a valuable gene and its sponsor are identified, introduced into the host species and then tested to ensure a stable transgenic heritage. Some commercially important species are derived from other animal genes. Coho salmon, for example, grew 11 times (0–37 times the effect range) as quickly as non-transgenic salmon with growth hormone gene and sockeye salmon promoter.

The private industry supports studies on transgenic fish and transgender Atlantic salmon, which develops 400 percent faster than average in the first year at an inland aquaculture facility in Scotland (Elliot Entis, NF Protein, pers. comm.). Long-term approaches to the development of the breed include the collection of long-term records, record keeping, monitoring and recording of the breeding stock. Specialized breeding facilities may be required to keep track of prospects, tools or opportunities to undertake those systems, for example during the early stages of the gene transfer system, or for all farms or territories (Zargar, 2020).

4.4 SHORT-TERM STRATEGIES

And some short-term genetic engineering experiments are similar to some long-term projects and changes that lead to major improvements in basic science over a short period. Although technology is easy to handle and sustain employee exploitation, it is not important for anyone to treat technology as "simple" or for proper management and record-keeping of the workforce, but rather for them to maintain safe, genetically appropriate, manipulative stocks.

The researcher decided to combine two genetically distinct groups in such a way that they would achieve the optimal mix of characteristics. This happens when they hybridize (heterosis). It is most likely that fish populations will also hybridize to larger fish, that is, if they do not release viable hybrid offspring or if they do not release these hybrid offspring into the wild.

Only the bodies of those born from the generation of vigor are spared, so it is necessary to preserve and treat their pure breeding lines. Many of the main aquaculture groups in which we are involved, such as salmonids and penaeid shrimp, do not yield useful hybrids. If a number of marine hybrids are produced, the success of their community can be and will be viewed as a metric. The Kuwait Institute of Scientific Research has found a fish called *Acanthopagus latus* (Sheim), named after the Arab poet Ali Shem. This fish is currently under investigation (Zargar, 2020).

It has a very high consistency and a solid body structure that also looks very desirable (Al-Abdul-Elah, pers. comm.). In the estimation, it is important to consider whether a cross has been made between humans and animals. At the end of the hook, the Sun Bass is a more successful predator than any reciprocal Hybrid in which X is hybridized with the wild Palmetto Bass, the Osteogloss. Thai catfish, *Clarius macrocephalus* and hybrid *Clarius batrachus* are chosen as females in Thailand (Mr Sujin, pers. comm). There are several types of marine hybrids, and their parent lines may be combined or even interbred. In Venezuela, the hybrid red tilapia is intended to encourage the development and shape of the body (FAO, pub. report).

Until DNA modifications continue to occur, different marine organisms (e.g. epigenetics) use thermal and chemical shocks to activate or deactivate any of their chromosomes (polyploidisation). In particular, the three species described in the Triple Sulfate are interesting because they can be sterile and thus give more energy to the process of growth than to the process of maturation and reproduction.

While not all species produce sex at all times, or even in most, other species do not produce sex at all. Some of them may not even contain triploids. The case of triploid species tends to be more common than that of diploid species. As in previous studies, there was an inconsistent development of triploids with Atlantic salmon; in Coho Salmon, the growth and survival of the triploids were worse. Oyster farmers have much more productive oysters (it is not common for oysters to multiply evenly) and oysters have a stronger flavor that is more suited to what consumers will pay for.

One of the noteworthy studies of hermaphrodite oysters was that of Guo, who had higher survival and size than triploids developed by chemical shock and were larger than diploid controls in triploid oysters. To establish cost advantages, the value of sexual dimorphism with substantial characteristics or where a reduced reproductive probability is required. E.g., tilapia is preferred to male and female rainbow trout is preferred to females because tilapia in males matures faster than somatoproducts and rainbow trout in females is preferred due to faster growth and slower maturation. The creation of more female groups would lead to more traditional fishing, such as sturgeon caviar. If you wish to prevent a sex change in fish, you should have an androgen environment and an estrogenic environment (by using fish of only one sex). While early use of hormones in aquatic animal production is widely recognized due to the lack of trace levels before the body hits the market volume, the use of hormone-treated fish is still forbidden in some countries. This dilemma can be solved by mating male vs. female fish in such a manner that they produce an entire female brood. It would prevent the use of hormones in eggs and sperm environments (Ray, 2018).

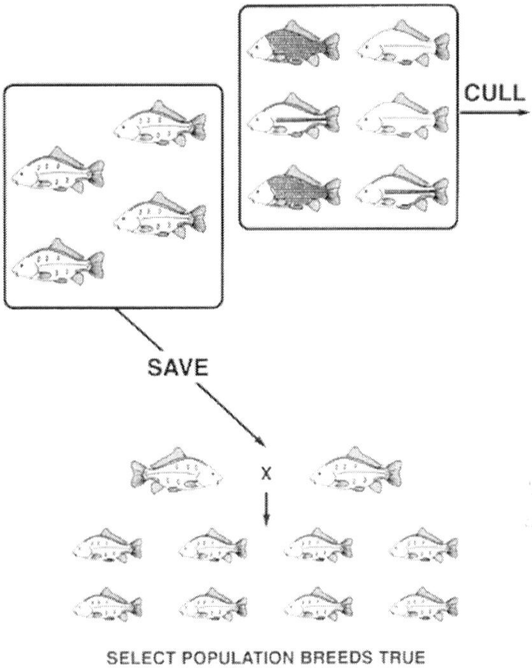

Figure 4.2: Selective breeding programs or medium-sized fish farm

Source: http://www.fao.org/3/v8720e/V8720E03.htm

4.5 COMBINING TECHNIQUES

When trying to modify a palindrome, several variations of techniques may be tried. A selectively replicated hybrid would begin by adding one allele that will perform very well under such conditions, while the other allele, which was mostly eliminated in natural forms, would become more dominant. According to experts, Regulation 26/2003/EC does not allow the use of GM crops in this method of the breeding process. From a scientific point of view, hybridization could increase the survival rate of hybrids while maintaining the fish that will be hybridized. Genetic engineering has been incorporated in the fish food industry. In the case of European Atlantic Salmon X trout crosses, sexual hybridization and polyploidisation have been used to strengthen cancer. Strangely, some triploid species of Atlantic Cod, Salmon Salar, and X European Foreheads have improved their survival rate and growth to the same degree as Atlantics. Triploid Pacific salmon hybrids were formed earlier during acclimatization studies. This fish displayed acceleration in disease resistance and growth rates were slow, but the reverse was seen with some hybrids. Fish conservation can be analyzed by using the genetics of many species of fish to gage the composition of the population, as well as by using genetics to see how the stocks are related to land and wildlife. E.g., Enriquez looked at the mitochondrial DNA (mtDNA) variety Brown Trout in 1996 and Barbosa researched isozymes in 1994 to decide that Atlantic Brown Truffle populations in the Orb River Basin were not interred in the Mediterranean Sea, possibly due to their lack of maturity in Atlantic populations in South France (stocked from hatcheries). While it was not considered a specific habitat that did not allow for direct genetic production, Atlantic salmon growth rates and return home were selected from Iceland and various stream release strains were evaluated.

4.6 GENETIC TECHNOLOGIES AND CONSERVATION

Farming in the sea when the atmosphere is natural is a good idea. Moreover, it is not without debate, and it is also expensive. Several organizations have described aquaculture as a significant challenge to natural habitats and marine biological diversity by releasing toxins (feeds, drugs, etc.) and mixing, for example, with farmed fish and wild animals. The treatment will also have side effects that are not entirely negative, as they are also helpful. The biggest concern concerning a genetically modified (GM) product is that it has new human and non-human features. And they would depend on the new host and therefore need to be more complex for the creation of new species in these new ecosystems of transgenic organisms. The reproducibility and viability of genetically engineered fish in the kind of hybrid or triploid species is both confusing, but their use has not led to the controversy of transgenic organisms. Biotechnology can be used to resolve a variety of environmental concerns in the aquatic food industry, including aquaculture. When it comes to salmon farming, the difficult interbreeding between livestock and hatcheries is a big problem for the company (Overturf, 2009).

While studies indicate that direct genetic inflows of wild animals are not feasible and are largely reliant on the lack of proper management of wild wildlife, it is difficult to identify the whole inflow of genetic material from wild communities, and a great deal of genetic material is caused by insufficient care of the aquatic environment. The most common species in aquaculture today is one that has either recently been eliminated from the wild or has only recently been imported into the wild. However, raising non-reproductive species would not minimize the scope for breeding aquaculture stocks from native plants, which added to futile requirements. Some examples are mono-sex, hybrid, or triploid varieties of non-productive organisms. However, there was the general belief that hatcheries, as well as hatcheries, would appear to be detrimental to the genetic makeup of wildlife (Overturf, 2009).

4.7 TOWARDS SUSTAINABLE AQUACULTURE

Projects for fish farming need more distinct genetic strains for aquaculture, as developed by selective breeding programs, and the basic task of raising fish for farming started as domestication began. One of the biggest issues facing e-cigarette industries is the transition of the market from a nicotine-only sector to an e-nicotine industry. There may be a few plants in cultivation that are split up into various categories of plants. Or how, by outbreeding, are many human animals domesticated on a local basis? This dilemma has come to the fore in the northern part of the United States with species such as Arctic grayling, winter flounder and Atlantic salmon. Will the selective breeding or gene transfer of the AFP antifreeze gene be tried, or are there any efforts that could be made to tame the winter flounce that can create these cold waters to nudge the limit in which Atlantic salmon will be raised? The International Conference of Aquaculture Gene Specialists addressed whether tilapia can be grown more successfully in saltwater. This was preceded by (saltwater cultivation for tilapia) which may apply to other contexts. These guidelines highlight a variety of topics, including studies that concentrate on local conditions, opportunities and dangers, as well as environmental and sustainability issues. It is believed that the relationship between genotype X ecology, i.e. the various selective pressures associated with different areas and agricultural systems, the development of agricultural diversity and the growth of locally adapted strains, are among the scenarios for diversification of aquaculture species. Interactions with the Genotype X ecosystem have long been considered unimportant in the production of genetically engineered tilapia as a commonly used example of genetic modification programs (Overturf, 2009).

Reporting can vary depending on the number of environments and genotypes tested. Scientists have genetically modified tilapia to make fish that contain lower amounts of toxin. Some conditions, such as underwater crevices, are part of a natural ecosystem in some areas and can, in turn, rely on unique genetic elements, such as genotypes in artificial ecosystems, and maybe also characteristics in brackish or aquatic settings other than those in freshwater ones. The lack of a well-designed method to assess the efficacy of breeding programs in diverse cultures makes it impossible to make a judgment on the success of these breeding programs. The introduction of genetics to the aquaculture industry has lately been discussed about environmental concerns, about the creation of genetically modified species, also known as genetically modified organisms (GMOs). The Canadian aquaculture industry has been called upon to cultivate non-reproductive salmonids (e.g. Atlantic salmon) to reduce the likelihood of releasing salmon from indigenous stocks. One way to do this is to make triploid salmon. These polyploid salmon have more disorders and frailty than haploid salmon, but they have a longer lifetime. The salmon farming industry does not want to harvest these salmon and has asked native salmonids to prove that they are facing a big challenge. First, the International Council for the Exploratory Protection of Marine and Environmental Modification Anthropology (ICES) has claimed, based on the genetic enhancement of C/I that Mediterranean terrestrial plants and animals that are easier to use have been using for years and this is the basis for the new creation/creation of the aquatic resource, the Intern (ICES). There are many biotechnological developments, such as hybridization and chromosome shifts, that farmers support, but biotechnological advances, such as genetic engineering, plant cell lines, or the use of transgenic animals, are more heated than others, such as conventional hybridization. In the past, scientists have said that advanced technologies can be built over a certain amount of time. Efforts throughout Europe, in both pure and commercial aquaculture, were to be coordinated by the European Aquaculture Society to optimize biodiversity and quality standards. All-inclusive topics in both the commercial and non-commercial world have been analyzed, and contingency plans for future activities have been discussed (Overturf, 2009).

These sexual organisms, such as monogamous and sterile individuals, make it possible for farmers and others to buy more eco-friendly and cheaply-grown fish collectives.

The SELAM network has worked very hard to build marketing campaigns that open up new streams of revenue and to ensure that the network has new services available, that they have diversified their goods, that they now have special relationships with nurseries and also with the fishing industry. The aquaculture industry is generally safe from outside interference, but the sustainability of this industry depends primarily on market interest and the acceptance of farmed goods. The assertion that genetically engineered plants cannot be poisonous is contradicted by the environmental groups who might condemn the industry as a whole. This will improve the reputation of the entire aquaculture industry by advancing genetic technologies for customer and environmental safety and discourage problematic (higher degree of uncertainty) or high-risk activities in areas of concern such as the release of fertile and poorly regulated GMOs. Heading to the molecular farming industry, the aquaculture industry should apply the same guidelines as agriculture to genetically modify plants.

Environmentalists and other citizens' groups actively argue that people should not eat genetically engineered food crops. A new study revealed that Unilever, a german food corporation, had shunned the sale of American Genetically Modified Maize (GMO maize) due to opposition from European Union officials. Several European countries have considered genetically modified crops and are working on new rules for the rest of the world (this is opposed by the USA and others). Our knowledge of the precise magnitude of the risks inherent in the processing of food from animals is not well known. We know not yet sure how to label certain foods since the EU has not defined how the label should be appended. The use of genetic techniques in aquaculture is a difficult topic due to religious and ethical concerns. The person concerned may have strong extreme feelings of distress in a case where the transgenic is still alive, but the degree of almost equal distress that the entity experiences may bring the transgenic animal too close to the human genetic norm to be acceptable for certain animal rights activists (DeSalle & Yudell, 2020).

4.8 FISH BREEDING

Natural and artificial selection

When attempting to cultivate a certain species of plant, one solution is to select a better plant from a natural organism that is not traditionally used. During early childhood, mutations that develop in long-term natural settings are chosen for animals. Their technological programs have been successfully extended. Interbreeding can lead to multiple differences between individuals, with birth defects being a more frequent outcome. There are also new forms of life that arise from these species, based on different combinations. This is 'domestication,' which is characteristic of newly developed species. This new race or variant would take a long history of evolution to see if there are any natural mutations. Maybe we should think about how "socially" we feel and change our lifestyle without altering our genes, and one day we'll look back at each other, "Seen you at the same table, same room, same time, same environment, same problem, different numbers, different physiology" The consequences of 'domestic adoption' are referred to as the 'psychotic effect.' In certain cases, the organisms may modify their gene at a low frequency, which means a form of "mutagenesis" that is associated with a type of weed or mutant phenoid. As these races originated from distinct areas in the same environment, with different pseudotypes or chromosomes, it was expected that they would have a long-term effect on different terrain, temperature, food and other uneven factors, such as a species subject to climate change. It was created through a long-term "natural" method, which was also generated through the use of mutagenesis, resulting in a clear imitation of the natural process. This describes its good home acceptance and capacity to mutagenesis species. It will be much harder to attempt to reproduce the process or try to find the factors that might make up something new and distinct from what is understood in our society. As there is only one planet and since farmers can grow across land and colonize new land to produce more crops, more farmers and more livestock, these animals must be defined and tracked. Sexual hybridization is the most useful option in the study of sexual hybridization (Bert, 2007).

Figure 4.3: Fish farming

Source: http://cascadebusnews.com/how-to-start-a-fish-farming-business/

Artificial sexual hybridization

In traditional breeding approaches, the breeding of new species involves genetic hybridization and/or sexual selection. Animal specimens can be tested to determine the characteristics of the organisms. To classify the optimal traits, one can either select from what is already known about the animal or see what is most different from other species. Because they may be developed by hybridization, fresh hybrids may be used as a form of genetic modification that will allow certain hybrid varieties to grow with better characteristics than their parents, which is what allows their offspring to evolve and become more productive plants. Thanks to frequent sexual hybridization, the couple has sex every day. None of these tests has yet been performed, of course. The idea of combining man-made techniques from males and females into a form of artificial DNA is to create a zygote, a cell dominated by double-stranded DNA that gives rise to the genetic material of the organism. Since the development of hybrid zygotes, they have a chance to grow into hybrids with certain enhanced properties. The process of piggybacking two sets of genes from two different genetically distinct chromosomes and converting them into a composite of two new genetic sets. The method is termed "recombinant homologous recombination" and the result is often referred to as "hybridization." Improvements in the morphological character of these hybrids have been explained by the influence of hybrid expressions of the newly reconstructed animal genomes. Hybrid vigor may be created by these (F1) hybrids, where closely related varieties of the species are chosen to represent the genetic features of the sexual hybridization involved. They provide, therefore, the basis for a new generation to be created by the precepts of Mendel's legacy. When the first variant characters differentiate male and female species, and the offspring of these 1st generation cross-breeds are mated together, the F2 generation offspring generation will represent the 1st variant, character family. Mutant animals can be created by mating different types of animals together. This method of hybridization is highly successful (Ray, 2018).

It would not be considered to be an effective method of generating 'very stable new animal species or forms.' However, it would not be considered to be an integral evolutionary mechanism since an acceptable evolutionary justification for this process is yet to be found. Moreover, since the different animal populations are not fully biologically compatible with each other, such sexual hybridization can only be accomplished between distinct species closely analogous to taxonomy. Cross-breeding between species takes place, however, fertilized eggs (zygotes) are only formed from male and female gametes. In rare cases, individuals can produce hybrid eggs that simultaneously contain both cytoplasm and egg nucleus. These eggs are fully sterile and cannot form the cytoplasm or eggs. They are completely grown, but they cannot extend as well, unlike normal egg cells. In virtually any large animal, from the amphibian to the mammalian class, it could be seen that this principle was being used. Studying suicide shows that one of the most common ways to commit suicide-jumping from heights -is by sexual hybridization between a male and a female gorilla. He had some solid characters from both his parents, but the project was unsuccessful. There are differences for lower animals, although there is no difference for higher vertebrates. Scientists have shown that sexual hybridization between closely related fish can be accomplished and occurs more often between some groups of fish, such as fish of different species on other planets. Given several types of "species" (like those described above), as well as subspecies, families, and orders, have been discovered (Ray, 2018).

4.9 APPLICATION OF GENETIC TECHNOLOGIES IN AQUACULTURE PRODUCTION

- Throughout history, humans have been made to eat a wide variety of marine invertebrates, mostly mollusks and crustaceans, dating back millions of years, and have also had and still have access to aquatic plants, especially algae, for food. With more than 8 percent of the world's population living in aquaculture and fisheries, and almost 50 percent of our water supply comes from aquaculture and fisheries, it is clear that the global economy is in a relatively stable situation. There is a great diversity of species. There are more than 31,000 species of finfish, 85,000 mollusks, and 47,000 crustaceans, 13,000 species of algae, 5,000 wild species, and more than 400 aquaculture species. As the economies of many nations continue to rely on fisheries and aquaculture, the monitoring and protection of freshwater and marine products need to focus on maximizing benefits from aquatic populations while protecting the integrity of marine and aquatic environments.
- A relatively new field of genetic engineering is taking form in the food industry. We don't have advanced software yet, but we're also working to improve the productivity. Genetics concepts have been applied to aquaculture plants. Genetic advances and the use of biotechnology have, in effect, been widely recognized over the last two decades as playing a very important role in the growth of aquaculture and, as a result of the proper execution of the proposed aquatic species genetic breeding projects, very important genetic profits can be made.

- The accompanying text of the COFI Working Paper: AQ/VI/2012/9, Genetic Opportunities and Advances in Aquaculture Growth: Possibilities and Challenges, which outlines current and potential uses of long-term and short-term genetic technologies for aquaculture development.
- Some types of genetic technology may be used in aquaculture to sustain the seed. But the scale is instead an increased yield in the primary. Genetic modification can enhance the ability of plants to survive under a wide variety of conditions. Furthermore, the expertise of genetic engineering scientists in a business can help to sell genetically modified products. It is argued that genetic engineering can be implemented for short or long-term benefit. Since short-term earnings taxes are typically immediate throughout their lifetime, they will usually not accumulate from one generation to the next. Long-term strategies, such as selective breeding, provide long-term benefits over the life cycles of many generations. A long-term approach to genetic enhancement.
- Only marine biology specialists are needed to understand the domestication and use of fish and other marine genetic tools. As far as domesticated and genetically modified strains are concerned, our agriculture and livestock sectors are still very much behind the aquatic and livestock industries.
- Development rates have been reported to be the most frequently enhanced feature of selective breeding systems, up to 20% per generation. The small number of genetic variations in the population is known to have other characteristics and their improvement is expected to increase. Using selective breeding methods, characteristics such as disease tolerance and stress can be continuously enhanced in terms of ripening time and quality of fresh meat. With anticipated and scheduled breeding cycles, new programs are now being established to support the production of new genomes. Examples of the varieties of fish used in recent breeding efforts include Atlantic cod, Atlantic salmon, common carp, golden sea bream, hybrid bass, Malawi tilapia lagoon, Mediterranean bass, Nile tilapia, red sea bass and rohu carp.
- The use of *Penaeus vannamei*, domesticated and genetically modified, white leg shrimp, for disease tolerance, has contributed to a substantial rise in the growth of shrimp aquaculture, but also

to a considerable risk of persistent infections, including viral pathogens that can be spread from the brood to the post-larvae. The main concern is that normally domesticated shrimps are used without caution so that the pathogen is not passed on to other shrimps.

- There are many ways in which genetically modified animals, such as carnivorous fish, can be selected as the primary energy source for products to make them easier to eat. One of them is the genetic selection or enhancement of the feed use of corn fish.

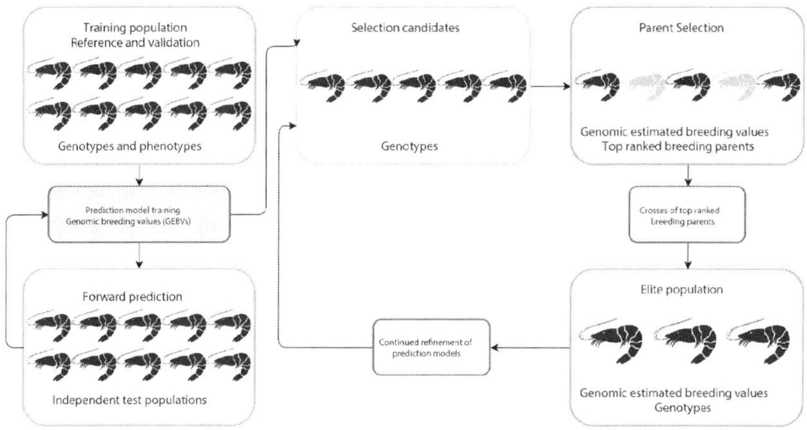

Figure 4.4: Genomic selection in aquaculture

Source: https://www.frontiersin.org/articles/10.3389/fgene.2018.00693/full

- The key driving force of the global fish aquaculture sector will continue to be the successful genetic modification and breeding of the best species for it.
- Aquaculture feed production, genetic engineering technology is now starting to help minimize dependency on fish meal and fish oil and thereby improve feed and feed ingredients such as real fish and plant-derived protein. Research with GM yeasts is being conducted for the development of main feed components such as carotenoid pigments and growth hormones in fish.
- Methods that can be used for pre-processing to minimize the impact of anti-nutrition.
- Plants can enhance their amino acid profile, which is linked to ant-dietary causes.

- A transamination reaction to convert low-grade animal by-products into high-value proteins.
- Transgenic fish (including puffer fish, discus fish, tilapia and carp) have been developed since the mid-1980s. They have been genetically modified to develop growth hormones. Patients usually experienced major improvements in development for several days before any development was made for several weeks. At this time, transgenic fish that are not released for human food/ biotechnology is not authorized.
- Genetic enhancement projects do not need to go through any regulatory filings or permitting processes. Almost all of this kind of project can be done in a short period, and the resulting artificially enhanced individual can have some very important changes or innovations in his or her life. This is a quick way to make it very effective.
- Conjunctions and intermediates will bring together two different types of seed and facilitate changes that will bring out the best possible characteristics (heterosis). The process of man-made genetic modification has played an important role in making the animals grow faster, eliminating the different genders of the animals, making the meat of the cattle look cleaner and also making the plants more tolerant of pest insects and so on. Hybridization with other species.
- Certain aquatic organisms have been exposed to thermal shocks and chemical forces that have altered either the number or the segregation of chromosome sets (polyploidisation). Triploid cells are not able to mature that well yet, which causes them to spend a lot of resources to develop while consuming fewer resources to reproduce. They're quite beneficial. Until now, a significant number of plants have produced triploids, but how useful they are will not be known until it is seen whether they affect human biology in the same way that proteins found in fish are known to have a profound effect on human biology. The use of triploids, such as triploid pacific oysters, has seen a significant increase in diploid controls from approximately 14 to 159 percent. While sterilization may not work on some species, some species are less likely to survive, on the other hand, posing significant problems with some projects, such as avoiding invasive fish and aquatic weed problems.

- Sexual enhancement means that cattle with extreme dimorphism, or where the chances are severely reduced, can be used to provide an improved reproductive rate. Gender-specific monosex male stocks have significant market advantages in a variety of species, particularly tilapia, due to both early maturation and unintended reproductive problems during the development period of this species. Some female fish, such as salmon, are also growing higher. Caviar also gives sturgeon to women. Government agencies that have issued strict regulations on trawling hormones have benefited many of the organizations involved in fishing (particularly in developing countries where trawlers are not widely available). However, many questions regarding the use of hormones in animal husbandry have been raised as a result of their use abroad, and whether this can be done in developed countries where medicines are marketed to export markets.
- Advances in genetic engineering, including the creation of genetically engineered species is just starting to be implemented. Recent genetic advances exploring how genes currently operated in organisms are used include DNA markers, new sequences, gene discoveries, genome mapping (revealing relative positions in genes that are observed in a single chromosome) and gene expression techniques. The changes that we can make in traditional aquaculture will help us identify some of the key biological factors, such as higher productivity, which will allow us to better select fish more quickly in the traditional aquaculture we have.
- Various major aquaculture products include the discovery of useful genetic tools for genetic development projects, the control of farmed health and the separation of natural and cultivated specimens, the analytical effects of escaped aquaculture on wild and genetic technologies for the classification of genetic resources.
- As the amount of information on genetics becomes more widespread, genetic markers will become more valuable in agricultural and industrial fields. Traceability is a key factor that needs to be considered in aquaculture certification schemes. Using GM technology, the presence of fish in one sample can be identified with great sensitivity, including frozen matter, filets, eggs and larvae, as well as early life history stadiums, which are

otherwise unidentifiable. Laboratories that are authorized to test fish have detected mislabeling and frauds of fish meds and have set up laboratories to help with the traceability of fish and meds.

- Pathogen detection using molecular techniques and immunodiagnosis is used extensively in the diagnosis of viruses, pathogens and diseases. Their advanced technical characteristics, sensitivity, precision and rapid diagnosis have played an important role in the management of health.
- The relationship between genetic programs that improve fishing and aquaculture techniques will improve the efficiency and quality of these programs in the future. It is important to bear in mind that there is a high level of skill and the need for special equipment involved in both of these processes.

CHAPTER 5

MOLECULAR MARKERS IN FISH

CONTENTS

5.1 INTRODUCTION

Molecular markers have been applied in fisheries. These are pecial fragments of DNA that can be found in the whole genome. Various sites are found in the molecular markers of the genome. The location of a certain gene or the ancestry of a certain phenotype is 'flagged.' Features of interest usually remain correlated with genetic cross molecular markers. Individuals where the molecular marker is present will be chosen as the marker suggests that the desired phenotype occurs. Various sites of molecular genome markers are established.

The location of a certain gene or lineage is used for the 'flagging' of a certain feature. There are different types of molecular markers used in fisheries all used for different purposes such as: Human recognition, genetic diversity and population composition study of natural species, distinction between wild and hatchery populations, demographic detection of bottlenecks and propagation support mechanisms for regeneration. The use of molecular markers will recognize and change genetic variations around the human genome.

This permitted rapid progress to be made in the study of genetic variation and inbreeding, the determination of parentage, the identification of organisms, the development of genetic linkage maps for aquaculture animals and the identification of quantitative traits associated with specific marker-assisted selection characteristics to be controlled using molecular markers in fish (Carvalho &Pitcher, 2012).

One of the molecular marker classifications refers to markers of type I as known feature genes and to markers of type II as unknown genomic segments. Markers with type I markers are well recognized. Form I DNA markers were originally not thought to be useful for fish genetic research, but over time it became clear that they were very important to wild and aquaculture studies. This marker has therefore become very important for the study of linkage phenomena and the mapping of QTL, which is of considerable significance for comparative genomic studies and for the description of candidate genes for quantitative properties in various aquaculture-related organisms. A significant number of experiments have been carried out in order to understand the genetic basis of the economically important features of different salmonid organisms. For example, in Atlantic salmon, QTL was mapped to loci associated with viral disease resistance, flesh color and growth characteristics, salinity tolerance, late maturity, etc (Carvalho & Pitcher, 2012).

5.2 ALLOZYME MARKERS

Allozymes are allelic versions of the encoded structural genes. Enzymes are proteins made up of amino acids, some of which are electrically charged. As a result, the enzymes have a net electrical charge that depends on the amino acid content of the protein. The net electrical charge of the protein may be modified and the full molecular structure (conformation) may be changed as the mutation in the DNA results in the substitution of amino acid. As the rate of protein migration in the electrical field can be affected by electrical charge and conformation shifts, gel electrophoresis and subsequent enzyme-specific stains can distinguish all variations. With an enzyme, it is usually possible to discriminate between two or more loci known to be isoloci. Allozyme variation is often commonly referred to as isozyme variation (Dunham, 2011).

Allozyme electrophoresis, a method for the separation of proteins of varying molecular sizes and electrical charges, is the simplest, more efficient and less expensive procedure for the identification of degrees of genetic variation between or within organisms. Resolutions are limited as DNA protein coding regions can only be tracked, and only a small proportion of improvements in certain regions can dramatically increase protein mobility. During migration, allozymes are visualized by chemical staining of the electrophoretic gel. The individual band positions in the homozygous gel and multiple heterozygous bands suggest mobility and the presence of allozygous variants. These multiple band positions represent allozymes, and thus gene frequencies, at the locations being studied.

In addition to their similarities to electrical load fluctuations, allozyme/isozyme studies include the identification of protein polymorphism Allozyme experiments and isozymes have become one of the most common approaches for researching fish genetic and stock diversity problems. They have been especially helpful in evaluating and analyzing biogeographic hypotheses for cryptic species. The advantage of Allozyme/isozyme electrophoresis is relatively simple, cost-effective and efficient. Another value is that the spectrum is distributed through a range of possible distinct gene loci. The main issue with the allozyme method is that the bands (all) with the same electrical load cannot be homologous and travel to the same point in the gel (i.e., evolutionary convergence) (Dunham, 2011).

The gel score is still very arbitrary since it is difficult to read when the bands are weak or close together. Differences have traditionally been considered selectively neutral, allowing the idea of separation times to be checked. However, certain studies have shown that some types of allozymes are not neutral and are not picked (Avise, 1994; Pogson et al., 1995; Powers & Shulte, 1996). In our view, increasing evidence suggests selective restrictions in various quantities for most (if not all) 'neutral' genetic markers, including allozymes, mtDNA and microsatellites. Consequently, the possibility of a collection of loci in the systematics does not preclude its significance. For example, morphologists also use characters that are the product of choice.

These markers have a variety of applications in fisheries. For instance, Allozymes, including measures of outcrossing rates, sub-composition and demographic divergence, were used in several population genetic studies. Allozymes are especially useful for research on crop diversity at the level of conspecific species and closely related animals. For fingerprinting and diversity study, interspecific associations, genetic inheritance and alloy frequencies in germplasm samples have been examined, often in conjunction with other markers, by serial processing in germplasm banks, and parents as hybrids have been established.

Allozyme markers are among the commonly used technologies in fisheries. This is because of the strengths they possess. For instance, the power of the allozymes is simplicity. They are short and easy to use as allozyme analysis does not require any DNA extraction or sequence data, primers or samples available. However, certain species may require significant technological optimization for some enzymes. Simple analytical techniques allow the integration of some allozymes at a relatively low cost, depending on the enzyme staining agent used. Codominant markers are allozymes that are highly reproducible. As far as loci and alleles are concerned, zymograms (the banding pattern of isozymes) may be quickly interpreted or it might be sufficient to explain the isolation of children from established parental crosses. However, zymograms also display complex polyploid or duplicating gene banding profiles and intergenic heterodimer production, which may make understanding more complicated (Overturf, 2009).

In as much as they have advantages, the use of allozymes has its own disadvantages. The biggest disadvantages of allozymes are their very poor abundance and low polymorphism. Proteins with equal electrophoretic mobility (co-migration) can also not be homologous to remotely bound germplasm. In addition, their limited neutrality can be questioned. In addition, allozymes are sometimes referred to as molecular markers, and since they represent enzyme variants, enzymes are molecules. Allozymes, though, are phenotypic receptors and can also be affected by environmental influences. For example, the banding profile obtained for a given allozyme marker may differ based on the form of tissue used in the analysis. This is since a tissue gene cannot be expressed in other tissues. On the other hand, molecular markers do not have an environmental effect because they depend on variations in the DNA sequence that suggest that the same banding profiles can be expected at any time for a single genotype.

5.3 RESTRICTION FRAGMENT LENGTH POLYMORPHISM (RFLP)

The length constraint of the polymorphism fragment is a type of polymorphism that occurs from the variation of the DNA sequence recognized by the restriction enzymes (RFLP). Scientists at known sites use bacterial enzymes to break down DNA molecules. RFLPs (pronounced 'rif lips') are used as markers for genomic maps. RFLPs are normally visualized using an electrophoresis gel.

The RFLP technique is a technique developed by the English scientist Alec Jeffreys (1984) during a genetic disorder experiment. It is used to study specific differences in DNA fragments in order to differentiate between genetically modified organisms. These patterns are called tandem repeat vector numbers (VNTRs). In more than 1% of the normal population, genetic polymorphism is characterized by hereditary genetic variations between individuals. RFLP uses these differences in DNA sequences to identify and analyze all intraspecies and interspecies variants. Selective endonucleases are enzymes that divide DNA into tiny parts for a long time. Each endonuclease restriction targets several nucleotide sequences in the DNA chain and cuts at different locations. The distance between a few endonuclease sites ranges from person to person. As a consequence, the length of the DNA fragments produced by the endonuclease restriction can differ in both human and animal species. A series of briefly mentioned steps are used to perform RFLP: DNA extraction. DNA is extracted and refined from blood, saliva and other samples (Shimizu et al.,2013).

DNA fragmentation; Filtered DNA is digested by restriction of endonucleases. In general, there are 4 to 6 essential pairs in length for the recognition of these enzymes. The shorter the known chain, the more fragments of digestion are generated. For example, if the GAGC chain is small, in the DNA sample. The endonu4clease restriction that recognizes the GAGC sequence decreases DNA as the GAGC pattern is replicated.

When a sample repeats the GAGC sequence four times and another sample repeats it twice, the duration of the fragment formed by the enzyme for both samples will be different. Gel Electrophoresis; During DNA fragmentation, the restriction fragments formed by the electrophoresis gel are analyzed. Fragments are negatively loaded and readily separated by electrophoresis, which separates molecules by size and load. Breaking DNA specimens like the electrophoretic gel and two electrodes are placed in the chamber. If the electrical field is applied, the fragments are forced towards the positive electrode. Smaller particles quickly migrate to the gel, leaving larger samples behind and separating the DNA samples into separate gel bands. Visualization of bands; the gel is treated with lighting dyes to make the DNA bands visible.

The study of genome-based RFLP variation was once a critical tool for genome mapping and genetic disease assessment. In order to determine the chromosomal status of a specific gene disorder, researchers will analyze the DNA in the patient's family to classify all RFLPs with an ancestry pattern similar to the disease pattern (see genetic linkage). In other families, RFLP analyzes indicate who was at risk of having a deficiency or who was possibly a carrier of gene mutations after clustering the disease gene. The RFLP test is used to identify and differentiate animals by studying the patterns of single genes. The rate of recombination between restrictive locations in the loci is also specified. Early genetic fingerprinting technologies were also developed in the RFLP research, which were useful for identifying samples from crime scenes, determining paternity and the presence of genetic variants or animal reproductive patterns.

There are other applications of RFLP in fisheries. They include RFLP being used in deciding on the diagnosis of a person's genetic disease, namely cystic fibrosis. For example, the source of the DNA sample can be determined or confirmed in paternity tests or forensic surveys. Identify the degree of recombination that indicates the disparity in the genetic mapping between the loci. This allows knowing the carrier of a family mutation that causes a disorder (Shimizu et al., 2013).

There are a number of disadvantages of RFLP technologies. Since its conception, RFLP has become a widely used tool for genomic analysis in forensics, pathology and genetic testing. However, DNA profiling methods, including polymerase chain reactions, are becoming increasingly fast and less expensive (PCR). The RFLP protocol requires many steps and takes weeks to achieve results, but techniques such as PCR can boost target DNA sequences in only a few hours. In addition, a large DNA sample, which may be long and laborious to separate, is needed for RFLP. On the other hand, the PCR can amplify a small volume of DNA within a few hours. For multiple applications such as this, the PCR technique has increasingly replaced RFLP in most applications involving DNA sequences, such as paternity testing or forensic sampling.

5.4 RANDOM AMPLIFIED POLYMORPHIC DNA (RAPD)

In these markers, DNA fragments for amplifying random PCR segments with a single arbitrary nuclear sequence are identifiers. The RAPD markers are applied using the RAPD technology. The accuracy of DNA obtained through different protocols and subject to different storage conditions can be tested by comparing PCR amplification profiles. The most common RAPD technique is available in almost any laboratory. Random polymorphic amplified DNA is a PCR-based diagnostic test for genetic variants (RAPD). This involves the use of a single random primer to amplify a variety of different DNA materials during a PCR reaction. The technique was developed separately by two different laboratories (Williams et al., 1990; Welsh & McClelland, 1990) (Arbitrary primed PCR). This method detects a polymorphic nucleotide sequence with just one nucleotide sequence primer in a DNA-based amplification examination. In this response, a single primary species binds to genomic DNA at two distinct positions in opposite strands of the DNA template. Thermocylic amplification produces a discreet DNA material as these primary sites are located at a distance that can be amplified from each other. Polymorphisms between entities occur at one or both primary binding sites, so differences in sequences are apparent as the presence or absence of a separate RAPD band. Thus, both polymorphisms are dominant genetic markers (Freeland, 2006).

This type of marker technology follows the following principle; the response of the polymerase chain is based on the RAPD framework (PCR). The target DNA sequence is exponentially amplified by random primers, thermostable DNA polymerase, phosphate, magnesium and buffer nucleotide reactions. The reaction involves repeated steps, each consisting of denaturation, primary rinsing and elongation. In the first step, the DNA temperature has risen to 94 °C (denaturation). The temperature drops to about 40 to 65 °C in the second half and the first sequence on the DNA template is flushed (annealing step). For the thermostable operation of Taq polymerase,

DNA is usually treated at 72°C at the third step. The thermostatic operation is chosen. Analysis was conducted to identify molecularly the most common species of fish in central India using Random Amplified Polymorphic DNA (RAPD). Research was done randomly from *Labeo rohita*, *Catla catla* and *Cirrina mrigala*, male and female animals from different parts of central Indian regions (family: Cyprinidae). 10 decanucleotide primers (RAn primers) were used for the polymerase chain reaction. Six randomized RAN1 (GATGACCGCC), RAN 3 (GGCACGTAAC), RAN 4 (GGCATGACCT), RAN 5 (GGGTAACGCC) and RAN9 (GTGCCAATG) primers were considered for genetic differentiation and band analysis based on interpretability, simplicity and reproducibility. Based on the series of random primers and genotypes used, various sizes of amplified products are available. There were a total of 55 discreet amplified pieces (size 200 to 2000 bp approximately). 29 of the 55 medications demonstrated strong polymorphism between species in 29 species-specific markers.

The highest and lowest RAPD band numbers for RAN 5 and RAN 9 were 12 and 6. The 16 families were the most important, as has been commonly seen in all three family species: For all three species of fish, both male and female, related band patterns have been collected, demonstrating 100% genetic sexual similarity. The DNA bands were distinct in nature and determined the genetic variation between the species of fish and the presence of common bands assigned to evolutionary relations. Special bands of species-specific patterns may be useful molecular recognition methods. The study restores the utility of RAPD-PCR molecular taxonomy studies in various species of fish such as tilapia (Freeland, 2006).

The population of Tilapia is one of the fastest-growing species of finfish farming in the world. It has broadened its base scope from subsistence-oriented technology to a global trade market as a high-quality filet sold to Europe and the US. Most Tilapian species focus on breeding, development, and feeding, structural and biogeographic characteristics. However, such characters are of no value for identifying purposes, since they exhibit major population similarities and small differences between distinct species. The use of electrophoresis proteins is common among tilapia, although this technique does not distinguish between Tilapia Nile subspecies. Mitochondrial DNA markers have been used to accurately identify subspecies. However, no attempt has been made to analyze nuclear DNA for this purpose. RAPD analysis was used to diagnose three species of Tilapia *Oreochromis* and four subspecies (Bardakci & Skibinski, 1994).

For different species, but not just for different sub-species, different RAPD fragment patterns have been established. Evidence has shown that systematic investigations of RAPD markers may be useful for species and subspecies. Nxomani et al. (1999) used RAPD analysis to determine color variations in *Tilapia guinasana*, on the other hand, in Northwest South Africa and Namibia. Data show the genetic distinction of the color of *Tilapia guinasana*.

The authors proposed a new solution to developing outbreeding in fully isolated areas, where the existing maximum is nothing more than a reduced capacity to carry out the production of sinkhole biome threats. Fish production in Egypt has grown significantly to 546,000 tons in 1998 compared to 306,000 tons in 1988 (26 percent). Work on fish genetics in Egypt has just started. Most fish farms are poly-grown, with tilapia contributing approximately 38% of total production. With tilapia, mullets and carp are also deposited. RAPD analysis examined four species of Tilapia (three genera): redbelly tilapia (*Tilapia zillii*), white tilapia (*Sarotherodon galilaeus*) and blue tilapia (*O. aureus*) *(O. niloticus*). Random 4-10-sea and 6-20-sea primers have been used for research on polymorphisms between three genera and between two species.

In the three gender zones, the total number of fragments produced per primer ranged from 3 to 15, while 16 of the 20 primers (84.21%) produced polymorphs. Results have shown that there are significant differences between the three varieties of tilapia fish. These studies have shown that RAPD markers can be useful for the systemic investigation of tilapia at the species level (Ahmed et al., 2004)

There are a number of limitations to using these kinds of markers. For instance, almost all RAPD markers are dominant, but they are invisible if the heterozygous (1) or homozygous locus amplifies the DNA component (2 copies). Watched as several fragments of DNA amplified from the same locus, co-dominant RAPD markers occur only rarely. PCR is an enzyme reaction, so the result of the DNA template, the concentration of the PCR component and the state of the PCR cycle can be determined by shape and concentration. RAPD is well known as a laboratory technique and must also be repeated with special laboratory protocols. The discrepancies between the first and the blueprint lead both to the complete absence of the PCR product and to a mere decrease in the quantity of the product. The results of the RAPD can also be difficult to read.

5.5 AMPLIFIED FRAGMENT LENGTH POLYMORPHISM (AFLP)

AFLP is a PCR-based multiplex technique where a subset of restriction fragments is selectively multiplied by the use of oligonucleotide primers to complement sequences connected to either end. The AFLP analysis makes it possible to measure more than 50 loci accurately in a single reaction. The reliability of RFLP and its simple PCR are incorporated into this technique; thus, AFLP is a modern DNA type system with a certain origin or sophistication. Numerous molecular markers are required to establish maps of genetic linkages and to pick a marker (MAS). Amplified polymorphism fragment length (AFLP) markers are derived from cats with Mendelian rules and are also useful for genetic mapping and MAS applications. AFLP pattern between *Ictalurus punctatus* and Catfish Channel was tested with 64 primer variants. A large range of AFLP markers is known. A total of 7,871 strips were multiplied by 64 major variants, of which 3,081 were polymorphic to the blue catfish used in mapping analysis for the development of reference families (Gupta, 2008).

Higher polymorphism rates are reported as primary combinations that produce fewer bands. The 53 primes generated simple reproducible AFLP profiles from the 64 primers. The Equivalent AFLP profiles are formed with nine primary variants. The other two primers provided low AFLP profiles with intense bands. In AFLP 2,572, 53 excellent first variations were made. These markers can be used to map genes and quantitative characteristics of catfish by applying an interspecific hybrid approach. Genetic variation analysis of channel catfish strains has demonstrated that AFLP can also be an important method for fish genetic and population analysis. A good example of the AFLP markers used in fish is the formation of an Ayu mapping (F1) family. It was formed when a normal diploid male was crossed with a homozygous clonal female. The genetic linkage was traced using 191 amplified polymorphic fragment length (AFLP) and 4 Microsoft DNA markers. A total of 168 loci were laid out in 36 linkage groups of approximately 1659.6 cm, comprising approximately 77.3 percent to 81.8 percent of the total genome. During the random dispersion of the markers around the genome, the development of broad mapping maps was very successful.

AFLP points, such as RAPDs, are dominant labels, but software packages are also available for co-dominant AFLP band scores (AFLP QuantaPro, Main Gene). Codominant scores are possible for well-defined families, but they are complicated for demographic research. This can be solved by experiments on the graphic representation of human incremental polymorphism (AFLP).

Improved application of the AFLP:

Step 1, Genomic DN A is digested into multiple fragments of various sizes by means of EcoRI and MseI; step 2, the adapters are added to the ends of the DNA fragment; step 3, the addition of a random base at the 3V end, which amplifies the 1/16-figure fragment; and step 4, the residual amplification; it is 14 Liu, J.F. Cordes/Aquaculture 238 (2004) 1-37 AFLP Gel Interest Bands to sequence, pla. This results in one or more co-dominant markers which, while time-consuming, can be graded with traditional analyses. A marker hybridization of native and imported tropics has been used in the California reservoirs. AFLPs may be used as dominant markers as a restricted response. When parents are homozygous, both maternal and paternal parents may be summed up in the F1 generation (Liu et al., 1998b, 1999c). The main strengths of the AFLP are the large (over 100) polymorphisms exposed, the high reproductively due to the high temperature of the PCR system and the relative economy per marker. The cost per marker is much more costly than the RAPDs, but large quantities from a single run can be analyzed. No previous molecular content, such as RAPDs, is essential and is therefore adequate for any specimen, including the less extensively studied species of fish. Like RAPDs, AFLP bands are also referred to as bi-allelic bands, with comparatively low PIC values, but the potency of these bands is greatly increased by more loci, which can be rated at the same time. Special equipment for electrophoretic analysis of fluorescent labeling is required, such as automated gene sequencer; traditional electrophoretic methods can also be used, but radioactive labeling or special staining techniques such as silver color are required (Gupta, 2008).

5.6 MICROSATELLITES MARKERS

Microsatellite markers are widely used in fish genetics, especially due to their easy-to-use and high-resolution functionality. However, there are significant prior genomic records of the research organisms and the use of microsatellites is restricted if the genetic studies do not contain any species-specific primers. It's been made recently. Owing to the presence of zero partners, the work of researchers has been hampered and the number of loci used has been reduced. Efforts to develop and use molecular markers to assess the genetic variation of species and to guide conservation efforts are also useful. Heterogeneous primers and the economic and ecological importance of these have been used. The purpose of this study was to define and classify B-specific primers for microsatellites for the crossbreeding of two equivalent species (*Brycon gouldingi* and *Brycon falcatus*).

Microsatellites are the most common and versatile genetic markers for a variety of uses in population genetics, conservation biology and evolutionary biology. DNA gene arrays contain replicated mono-, di-, tri-and tetra nucleotide units distributed across a variety of eukaryotic organism genomes. Microsatellites are of a strongly polymorphic and readily characterized genetic form. Both are very suited to the study of population structures and pedigrees and can discern distinctions between closely related species. PCR can be automated for microsatellites to detect repeated sequences of simple polymorphisms. Blood samples or alcohol-containing tissues must be tested in a limited amount. The majority of the microsatellites are not encrypted; hence, the discrepancy is independent of normal preferences. These features include a genetic marker for microsatellites for genetic security and fisheries management. This thesis explores the use of microsatellite markers in conservation genetics and recent developments in population structure research in fish management (Bert, 2007).

Satellites are made up of several thousand base pairs, repeated thousands or millions of times. Minisatellites are DNA sequences with a period of approximately 9-100 BP observed in a range of 2 to several hundred times. Minisatellites found in human insulin gene loci with repeat unit lengths from 10 to 64 bp have also been referred to as "variable tandem repeat numbers" (VNTRs) for DNA. Microsatellites with a fixed length of 1 to 6 bp are repeated up to 100 times in each locus. They are sometimes referred to as "simple sequence repeat" or "short tandem repeat" (STR) by DNA. The variations in the length of the replicate microsatellite and mini DNA is usually traitd to an increase or decrease in the number of replicate units, according to Jeffreys et al. and Weber. Differences in repeat numbers are the basis for many of the DNA profiling strategies used today. Later, only microsatellites are very common in genetic population studies.

Microsatellites are small repeat sequences of di, tri-or tetra-nucleotide arranged in tandem and repeat 1-6 bp repeatedly flanked by non-repeat special DNA sequence regions. For the first time, Tautz and Weber demonstrated polymorphism and can in microsatellite loci. Microsatellite loci can be used to measure genetic variations in or across animal populations by multiplying the polymerase chain reaction of small genomic DNA samples and alleles on a single and two-band polyacryamide gel separately and correctly. Microsatellite-based heterogeneity, processing speed and the ability to isolate large quantities of loci offer a marker mechanism that distinguishes between closely related populations. Single locus microsatellites, in which heterozygous expression is seen by both alleles, are commonly used in demographic research. The number of tandem repeats differs from the corresponding locus and, as such, electrophoresis can be defined on the basis of their dimensions (usually PAGE). There are a variety of repeat units that describe all of them in the locus. They have the same kind of specifics as allozymes: codominant alloy differentiating loci, but they are typically neutral and more variable than allozymes. A microsatellite is all genetic in Mendelian fashion, such as allozymes. In addition, alleles can also be consistently graded and compared over several gels (Bert, 2007).

Another benefit is that it is easy to use a minute or weakened DNA. Generally, there are several microsatellite loci distributed across the eukaryotic genome, and each locus has a known DNA sequence. The sequences are composed of single DNA patterns and several DNA patterns (which may be shared among loci). Single repeated tandem replications (SARs), typically consisting of 2 or 4 nucleotides (e.g. (AC)n) and (GATA)n) where n is between 5 and 50, consists of recurring components. Replicate dinucleotides—GT and CA—are the most abundant micro-sates in vertebrates. For the investigation of single locus microsatellites, similar primers flanking replication units, the sequence of which can be extracted from I DNA libraries, or (ii) from accessible gene bank sequences, are essential. These two methods are usually used to establish microsatellite markers. The second method is well established in the next chapter. In a study, Zane et al. showed many methods for the processing of microsatellite markers.

The key advantages of MSM are codominant distribution, nature-specific locus, highly polymorphic and hypervariable, broad patterns and details, relative abundance of uniform genome replication, higher than average mutation rate and rapid sampling. The advantages of micro-satellites, such as a shorter range, a continuous range of identical replicating units, high polymorphism, observations into mutation mechanisms, their occurrence in fish genomes, and the existence of the microsatellite cloning system, all have contributed to the growth of St. tetra-nucleotide microsatellite. The beneficial properties of microsatellites have been complemented by new technology such as automated capture and automated detection, and rating schemes such as automatic DNA sequence and fluorescent imaging. Microsatellite nuisance involves the presence of shadow or stutter bands, the frequency of zero alleles in some areas (non-observing alleles using standard assays) and homoplasticity. Microsatellite-flanked regions (MFRs) would also provide longitude variants that could contribute to identical longitude variant studies (and comparisons of variance levels for homologous loci between species) that could challenge population level stories for microsatellites and phylogenetic inferences, since such variable lengths in flanked regions could potentially minimize repeat all-length variations (Bert, 2007).

5.7 HIGH GENOME COVERGE MARKERS

The use of genotyping has enabled the classification, mapping and storage of genes, population genetics, development, ecological speciation and invasion and genomic analysis, sex control, sex determination, diet, disease biomarkers and quantitative loci mapping of marker-assisted fishing and aquaculture. Recent developments in DNA sequencing technologies over the past decade have contributed to the detection of high-performance variants in many species. Fresh, high-performance, reliable and cost-effective genotyping techniques are urgently required to optimize access to rich genetics in animals. This is known as genotyping-by-sequencing (GBS) technology, which is a very promising research technique due to its potential to simultaneously recognize and genotype low-cost markers and clear molecular biology workflows for fisheries and aquaculture studies. GBS has been introduced in more than 50 species/experiments by the end of 2014 after it was first formulated for rice in 2009. It is now increasingly used in fisheries and aquaculture and has been launched in nearly 40 species/studies from 2015 to the current. It outlines the genotyping methodologies, recent advances in DNA sequencing technology to achieve GBS over the next decade, and the commitments made in this approach as a Genomics platform for fisheries and aquaculture genotyping. It also explores the promise of full genome sequencing (WGS) in GBS and adds advances in fisheries and aquaculture (Shimizu et al., 2013).

There are a number of applications of these markers. Genetic technologies can be used in a variety of ways in aquaculture, but the main use is to improve yields. Marketability increases, disease immunity, body shape and color among others. Efficient genetic technology will help the potential to expand and preserve natural resources. Genetic engineering systems can be seen in the short and long term. Short-term effects are typically immediate over the past two decades and are not generally stable, but long-term strategies, such as selective breeding, benefit each generation. Changes in protein, RNA, and DNA are important and related to certain diseases in a variety of biological processes. Accurate identification of protein, RNA and DNA-modifying sites, as well as a thorough understanding of them, would also promote studies into treatment and disease prevention. The number of sequences found started to increase with the introduction of sequencing technology. Over the last decade, a variety of computer techniques have been developed to predict protein, RNA and DNA sites. In this study, we showed the intermediate site predictors in detail for three biological sequences and the relationship between diseases.

These genetic variations are crucial to the survival of species of fish ecosystems that enhance the ability of animals to adapt to changing conditions. There are genetic variations between individuals, which contribute to a distinction between population, species and higher-order between taxonomic groups. Genetic diversity reports have a broad variety of applications in evolutionary research, natural resource conservation and management, and genetic enhancement programs. Molecular genetic markers are developed with a high degree of ability to diagnose genetic studies of persons, communities or animals. These molecular signatures, along with recent analytical developments, have revolutionized the predictive potential to study genetic alteration. Molecular markers and their mathematical interpretation have revolutionized the computational capacity required to determine genetic variation (Shimizu et al., 2013).

5.8 SINGLE-NUCLEOTIDE POLYMORPHISM (SNPS)

Geographic extension of aquaculture zones, the use of new organisms for agriculture and the use of new technology would increase the productivity of aquatic animals. The use of genetics and genomics is one of the recent developments in aquaculture development. New molecular instruments to support aquaculture have been developed. Numerous scientific and review articles on molecular markers and the variety of applications, including aquaculture and food product monitoring, have been published. Single Nucleotide Polymorphism (SNP) analytics have been developed as a generation technique for large-scale aquaculture applications. PNSs can be used to create genetic mapping maps, to identify quantitative characteristic loci (QTL) for useful features such as growth, body weight, grilling, low tolerance heat and oxygen, stress and disease resistance, mapping loci in selective and chromosome modulation tests for sex determination, and genome-wide selection and marker-assisted selection. However, multiple traits are related and can be governed by the QTL volume. The genotype has emerged as an interesting and important selection technology through the sequencing of the genome reduction problem. The discovery of genes, sequences and nucleotides (substitutions) that specifically influence phenotypic variations gives one the ability to select suitable culturally supported characters. You will upgrade the NPS and QTL ties using genome editing methods. Examples of successful application of SPS in aquaculture ecosystems for fish, crustaceans and mollusks, covering most regions, and an overview of environmental issues are addressed.

SNPs markers are mostly applied in generating linkage maps. Linkage maps indicate the chromosome location of the genetic marker. Due to its size and development, PNA figures such as up to 96000 Atlantic salmon PNAs have been developed in high density and rainbow truffle. Linkage maps for male and female Atlantic salmon were constructed on the basis of the PNS. Differences were determined between the rate of recombination between men and women. The linkage chart was helpful on the physical map of the salmon genome. An optimized connection map with 14,620 SPS loci, 286 salmon and 11,728 EST scaffolds for Chinook salmon was created. Genome regions with additional demographic variations have been identified and the fitness-related candidate genes, such as stress, growth and behavior, have been identified. QTL coexists with other HSP70.0 candidate genes (Manfred Reinecke, Giacomo Zaccone, & B. G. Kapoor, 2020).

A high-density genetic association map for Ictalurus punctatus catfish with the genotyping of the families of the 250K SPS sequence has been developed and integrated into a physical map. More than 50,000 Mossads have been mounted on this map, comprising 90% (867) Mb of the catfish genome. The estimated genetic scale was 3505.4 cM and the resolution for the sexually averaged genetic map was 0.22 cM. The optimized map made it possible to assemble the catfish genome, to find QTL and to clone the genes responsible for market-beneficial traits. For hybrid catfish (*I. punctatus*) and blue catfish (*I. furcatus*) channels, a high-density genetic relationship map has been developed (Liu et al., 2016). More than 26 000 GSPs are mapped to twenty-nine connection classes. Important discrepancies were found between the expected Mendelian ratio and the two main genomic blocks, less than 1000. The high selection was suggested as an explanation against all female blue catfish.

A standard genetic carp map (*Cyprinus carpio*) with 8487 PNS from 50 associated gene coverage groups of 3762.88 cM was created. For the genomic DNA sample of 68 individuals, 217 microsatellites and 336 SPS markers were collected from F2 mirror carp hybrids. AUTHOR: (Li et al., 2011). A map of the genetic linkage was identified and 14 QTLs for body weight were identified. In the preliminary QTL mapping process, a backcross family of 86 offspring was used to create a genetic map. 14 QTLs of body weight, body length and condition factor have been shown to be associated in 10 linkage groups. A common carp was assigned to create a genetic map consisting of 50 linkage groups with a total of 978 microsatellites and 3899 SPS markers. Ten QTLs were identified with the diameter of the eye and 20 QTLs with the cross of the eye were identified. A thick NPS-based interconnection map was built with an interspecific cross between two closely related cyprinid species in Japan: *Gnathopogon elongatus* and G lake dwellings (Reinecke et al., 2020).

With 1622 RAD-tag points, the linking map contained 1390.7 cM with an intermediate gap of 0.87 cm. Multiplexed genotyping strategy for generating high-density genomic maps for orange spotted grouper, *Epinephelus* (You et al., 2013). A total of 4608 SPS, spanning 1581.7 cM, with a median distance of 0.34 cM between SPS, will be seen on the sex-average diagram. The SPS was used to establish physical maps and linkages for the yellowtail (*Seriola quinqueradiata*), an important fish aquaculture species in Japan (Aoki et al., 2015). A double-digested genetic association (ddRAD) mapping was developed for Japanese eel Anguilla japonica (Kai et al., 2014). The female chart was 1748, 8 cM, and the male's map was 1294, 5 cM. There was an anchor point with a total of 2672 SPS markers for 1252 scaffolds with an area of 151 Mb. Maps of genetic linkage, namely golden-headed sea bream, *Sparus aurata*; Asian sea bass. *Lates calcarifer* and Mandarin trough, *Siniperca chuatsi* and Turbot: *Scophthalmus maximus*.

Other applications of SNPs markers in fisheries include establishing growth traits, determining the sex of fish, fish resistance to pathogens and traceability of fish among others (Solé-Cava et al., 2013).

5.9 MITOCHONDRIAL DNA (MTDNA)

Variations of mitochondrial DNA (mtDNA) in more than 40 species of fish have been investigated. Research indicates that mitochondrial fish genomes display small variations in size, but genomes of closely related species are also different in size and suggest that the removal and incorporation of counterbalances help to control genome size. Genes are still stable in order and structure, but nuclear sequences show accelerated separation. There are a number of distinct mtDNA haplotypes and a range of their rare mutational derivatives in most fish species. In environmental conditions, the degree of genetic heterogeneity between dominant haplotypes appears to be greater than that of marine species, especially those occupying non-glacial regions. Haplotype frequencies are typically highly divergent geographically, largely due to stochastic fluctuations in population-based occurrences. The basis for this difference is both local segregations in stock and the recognition of significant genetic discontinuities that characterize past incidents of genetic differentiation. Hatchery's inventory analysis also showed differences in mtDNA, although stocks of haplotypes, which are common in the wild, are in most cases known. There is also a possibility for the development of rare haplotype-marked floodplains that can be detected over many generations of life (Beaumont et al., 2010).

This marker has been extensively applied in fish identification more so in tuna. *Thunnus* spp. is a Scombridae that comprises 8 species commonly referred to as tuna and is considered to be one of the most commercially valuable fish in the world. The most widely traded species on the international fisheries market are Atlantic bluefin tuna (*Thunnus thynnus*), bigeye tuna (*Thunnus obesus*) Southern bluefin tuna (*Thunnus maccoyi*), yellowfin tuna (*Thunnus albacars*) and the Pacific bluefin tuna (*Thunnus orientalis*). They are morphologically very similar to those expected for congenerous species, particularly in their exchanged forms it is difficult to distinguish between species. Various protocols for identifying marine product species have been developed, but the low genetic variation between many tuna species typically confuses what is achieved by the molecular marker used.

In addition, some researchers have found that the use of the DNA barcoding system is not productive and that all species cannot be classified due to the lack of resolution of this mitochondrial marker. The potential regions for the delimitation of species within the *Thunnus* genus have been identified here. In the mtDNA region between sites 8092 and 8847 Polymorphic sites for the proper delimitation of species containing the mtDNA genes ATP8, ATP6 and COX3 have been presented. The DNA barcode region also maintains the benefits of primary annealing relative to the fragment size (approximately 750 bp) and retained flanking regions. Samples obtained from local market fish are currently being assessed. Flanking primers have been developed (Beaumont et al., 2010).

As a marker, mtDNA has contributed to developments in fisheries. DNA barcode may be an important tool for species recognition based on partial mitochondrial cytochrome c Oxidase 1 gene sequence DNA barcoding (CO I). The method offers many advantages, such as the small number of biological samples needed, the applicability and the differentiation between phenotypically related species in all phases of life. Our research focuses on economically and ecologically important salmonid barcoding. Sequences have been determined by using special primers as the so-called universal primers struggle to produce amps. Sequences for BLAST species GenBank and BOLD databases are evidence of recognition. In one exception, the results of the barcoding were consistent with those based on morphology in cases in which the above diagnosis was made. The method can be carried out regularly for species identification, but it is entirely pointless when working with hybrids, and should be avoided with species that can hybridize themselves spontaneously (Hart & Reynolds, 2008).

CHAPTER 6

CATFISH HYBRID TECHNOLOGY

CONTENTS

6.1. INTRODUCTION

Innovation improvements have assumed a key job in the worldwide aquaculture improvement and increment in profitability. Catfish creation advances that heighten creation have progressed quickly in the new year's, yet not the sum total of what developments have been received promptly by ranchers. A new overview of the catfish makers in Alabama, Arkansas, and Mississippi uncovers some significant elements that impact innovation reception among catfish farmers. The fastest appropriation has been of the part lakes, with 1,700 sections of land in 2013, trailed by seriously circulated air through lakes (1,200 sections of land). High return and more noteworthy power over the creation cycle related with the other creation advances were the significant purposes behind innovation appropriation. High venture costs and expanded creation and promoting hazards related to these advances were the primary impediments for not receiving the new innovations (Dunham, 2011).

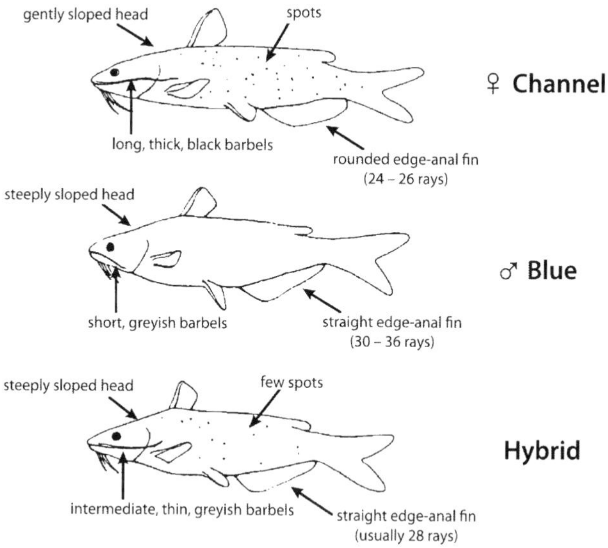

Figure 6.1. How to produce hybrid catfish?

Source: https://appliedecology.cals.ncsu.edu/wp-content/uploads/190.pdf

Creation of half breeds began in 1966 with two breeds of catfish that are catfish (*Ictalurus punctatus*) and blue catfish (*Ictalurus furcatus*). Examination to create and refine techniques for delivering cross breeds and to assess their exhibition for monetarily significant traits has proceeded until today. The hybrid for the most part performs in a way that is better than either parent species for a few significant traits including endurance, development, disease resistance and yield. The crossover is done by making a female catfish and blue male catfish to cross-breed and it is important to note that this is generally how normal hybrids also are created. Subsequently, all through this manual the term cross-breed alludes to the catfish female x blue catfish male cross-breed.

The essential imperative to business creation of the cross-breed is the absence of dependable, savvy techniques for delivering enormous amounts of fry required for business catfish cultivating. Notwithstanding, refinements of procedures for delivering half breeds and general predominance of crossovers comparative with channel catfish have prodded reestablished interest being used of half breeds for business creation..

In spite of the fact that there are reports of fruitful generating in brood lakes loaded with the catfish, the event of lake generating is extremely uncommon and not dependable for business incubation facility creation. Except if some future development drastically improves the spawning rate in conventional brood lakes, creation of fish will rely upon the utilization of hormones to activate ovulation (last development and arrival of eggs) (Dunham, 2011).

Hormone-induced creation of haploid catfish includes infusing female catfish with mixes to activate the ovulation process, deciding the event of ovulation, physically removing the eggs, and afterward mixing the eggs with sperms from the blue catfish. After treatment, the eggs can be brought forth utilizing methods used to incubate the catfish eggs. Techniques for hormone-induced creation of cross-breeds are grouped. In one, , developed catfish of both genders are put in a tank and the female catfish is infused with hormone (in some cases the male catfish is also infused however infusion of males is presumably redundant). Regularly the fish are put in an aquarium where they can be monitored even after the infusion process.

The catfish that is female is deprived of eggs that is treated with sperms from the blue catfish. Initially, when the male catfish is present, the conduct cooperation of the two catfish was important during the ovulation process to ensure proficiency and the approach was a regularly utilized strategy. In any case, research conducted at Auburn University showed when groups of female catfish when infused and kept in the aquarium, ovulation happened without the help of the male catfish. Gathering generating permits bigger quantities of females to be produced in a more proficient way. Subsequently, the methods depicted were identified with several spawning channel catfish females to deliver crossover fry. These include: great brood stock quality, legitimate handling and organization of hormone measurements, appropriate testicle assortment and sperm readiness, precise assurance of when ovulation would take place, great preparation strategies, and treatment of eggs.

Hormone-induced generating is an ordinarily utilized procedure for the creation of fry of many fish species and an assortment of mixes has been effectively utilized for prompting the ovulation process. These mixes create a chain of occasions that bring about the ovulation process and the release of eggs by the female catfish. CCP and LHRHa are mixed together and the mixture is usually used to aid the ovulation process and it has proven to be the best method so far (Dunham, 2011).

It is vital to note that when the two catfish mate they do not create a mixture that has the same properties as those in the CxB. Examination of C×B crossovers shows that they are hereditarily enhanced for a few economically alluring traits. This improved execution for some business characteristics in a solitary cross is uncommon, which renders the C×B cross-breed the best illustration of hereditary superiority to be accomplished in aquaculture with respect to general execution.

The C×B cross-breed is better than the normal catfish in light of the fact that of its: develops at a higher rate than the normal catfish, it feeds better, can function and survive in areas where the oxygen concentration is low, it has the ability to protect itself from diseases, resistance of packed development conditions in lakes, health and filet yield, and expanded harvest ability.

The presentation of C×B crossover is affirmed on business catfish ranches. Approximately two decades ago, C×B innovation was not monetarily doable. The gametes are viable, yet both catfish only sometimes interact with each other in light due to different variables. Nonetheless, ongoing advances in counterfeit generating and preparation procedures improve the fry creation and monetarily possible. The other significant improvement is the utilization of various strains of different catfish to make the crossover, which created hereditarily unmistakable C×B mixtures with much more predominant qualities (Dunham, 2011).

6.2 BROOD STOCK CARE/DETERMINATION

There are continuous exploration tasks to decide the impacts of diet, strain of fish, and other brood stock the executive's issues on the creation of half and half fry yet results are not indisputable. There are more seminars that have been conducted which aim at enhancing knowledge that is required for cross]-breeding the catfish. Hence the equivalent brood fish care and farming rules (taking care of, water quality and so on) initially, planned catfish production can be utilized to get ready catfish for hormone-induced bringing forth. There are changes that are supposed to be made to make this method effective and one of the changes is between the executives of brood fish for lake versus hormone-prompted spawning that will bring about stocking densities that are higher and those that can most likely be utilized and the extent of males can be diminished in the event that fish are just being utilized for hormone-induced spawning.

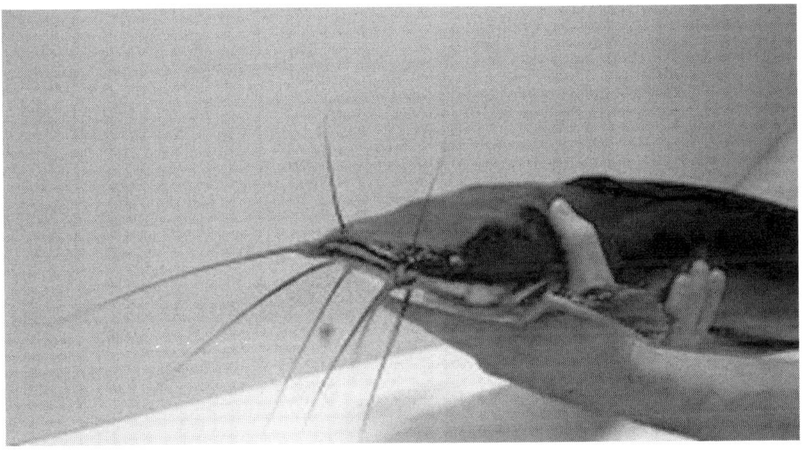

Figure 6.2. *Broodstock care in catfish hybrids.*

Source: https://www.google.com/url?sa=i&url=https%3A%2F%2Fwww.semanticscholar.org%2Fpaper%2FProduction-of-Hybrid-Catfish-Dunham

The presence of male catfish in the lakes is responsible for improving the genetic composition of females. This makes the utilization of such high densities to be the wrong choice when it comes to cross-breeding. However, it is vital to understand that not all female species are accepted for the infusion process. Approximately 1/3 of the female species are accepted to undergo the process. It is additionally essential to have a few stocks of brood fish in light of the fact that nonstop seining of a similar pool probably will pressure females (Biljana Vucelić-Radović, 2019).

Age is an important factor when it comes to cross-breeding because age determines how superior the hybrids will be and if they will have the best desirable qualities. The female catfish that should be used for the infusion process should be less than three years as they are considered to be more established. It is advisable to consider using young catfish for them to produce the best result. Consequently, it bodes well to utilize more established, bigger females for cross-breed creation. Size and weight is another factor that should be considered. The fish to be used should be approximately five to eight pounds. More smaller fish don't create numerous eggs per fish while stripping enormous quantities of huge females can prompt exhaustion.

Deciding on the right time when the female fish is prepared reproductively is the least demanding approach. When the fish is reproductively prepared it is hormone-prompted spawning they are put in a couple of jars. After this process, when the results are generated one can expect some part of the females prepared to begin producing. Another pointer that shows the fish are ready to reproduce is when they start making movements around the lake or in the ranch where they are taken care of. Quit taking care of around 3 days prior to choosing females for producing since one of the best pointers of their utility for producing. On the off chance that the fish have been as of late taken care of, it is hard to discern whether the totality of the midsection is because of the food in their stomachs or their ovaries which are large in size. In cases where the fish's needs are already fulfiled it is important for the person in charge of the process to discontinue the feeding process and ensure that the fish no longer receive their food then check on them after. For the selection process, it is easy to identify the females which appear to have swollen stomachs. Examining the female fish by holding them by the tail and then checking to see if they have swollen stomachs is the best method used to choose the fish (Vucelić-Radović, 2019).

Another way that can be used during the selection process is choosing female fish that have apertures that are swollen and red in color. Most people choose the fish by looking at their stomachs and choosing the ones with swollen ones. This method is preferred by many people because it is fast and easy. It can also be used to examine a large number of fish needed for cross-breeding. Fish with level or just somewhat swollen stomachs are for the most part poor contender for spawning and ought not to be utilized. The process can continue even after the fish have been chosen by conducting the same investigative process by checking the regions of the lakes that were not checked initially and fish that were not selected on the first round. When one does not receive satisfactory results it is important for them to seek another lake with other fishes to conduct their research. This is done to ensure that the female species that are used during the generating season are of high-quality because using fish that are of low quality is waste of funds. This is because the hormones that are to be injected into the female catfish are quite pricy and should only be used on the best female catfish.

Limiting pressure during the determination methodology is vital, keeping the fish from areas that do not have oxygen concentration away and if the climate is not cool the person in charge of the process should only conduct it when the water temperature is cool. The fish should not be used when they have already stayed out of the water for long r rather when they are not fresh. Handle the fish rapidly yet cautiously and load them onto transport tanks with satisfactory air circulation, utilize pure oxygen when conceivable. Abstain from harsh handling, since it could bring about harm to the ovary and lead to dying, what's more, coagulating which happens in the ovary.

Infusing females previously or after typical lake spawningaction may decrease crossover creation. It appears to be consistent that you would have better outcomes with hormone-actuated generating in the event that one is infusing fish in conditions where they are produce the best results. There are achievements that involve utilizing water wells that use warm water to build temperatures of the lake and quicken the development cycle with the goal that half and half creation can start sooner compared to conceivable dependent on cycles of temperature that are normal. Notwithstanding, the stream from warm water springs is for the most part restricted and significantly affecting the water temperature requires utilization of decently little lakes (Vucelić-Radović, 2019).

Blue catfish are considered to be immature till they are between four to five years. Blue catfish males 'by and large' have all around created testicles during a similar time period that female channel catfish are prepared to bring forth. Nonetheless, testicles improvement changes generally among individuals and is hard to anticipate an individual'stesticles improvement (and hence its convenience for half and half creation) in view of how the fish appears in terms of size and its general outward appearance. Utilizing the techniques that are considered to be current, how the testicles develop is not clear until after the male has been slaughtered and the testicles have been precisely isolated.

Improvement of a fast test to recognize how the testicles of the blue catfish develop in fish that are actually alive will be an added advantage in this process. There are certain procedures that can be used to determine whether the sperm is of good quality or not. Those species that are not used during the process can be kept in tanks or aquariums, what's more, utilized for the following cross breed creation cycle or got back to lakes (Bert, 2007).

6.3 HORMONE DOSAGE AND ADMINISTRATION

There are mixes that are used during the process and they include CCP and LHRHa. LHRHa stands for luteinizing hormone. The two mixes are successful and the sentiments on which is best for haploid creation fluctuate among users. We have utilized both effectively and both have favorable circumstances and inconveniences. LHRHa is a manufactured item and hence it is a setrile and reliable item. Likewise, the Investigational New Animal Drug which is abbreviated as INAD which can be acquired from different fish and wildlife found in the United States for LHRHa has the greatest limit of 100g/kg which is the standard weight for a female fish, which is by and large viewed as lower than the successful portion from the get-go in the spawning season. INAD is essential because it utilizes the two mixtures during the generative process. The mixes comprise of LHRHa and CCP. It is important to note that the time taken from the day of infusion to the day when ovulation begins is different in fish that used LHRHa and those that used CCP. The situation can either be an advantage or a disadvantage. The disadvantage is that time will be wasted when checking to see ovulation has begun to take place in both cases. On the other hand, the ovulation date for each fish will be more accurate and hence better results will be achieved (Andy Beaumont, Pierre Boudry, & Kathryn Hoare, 2010).

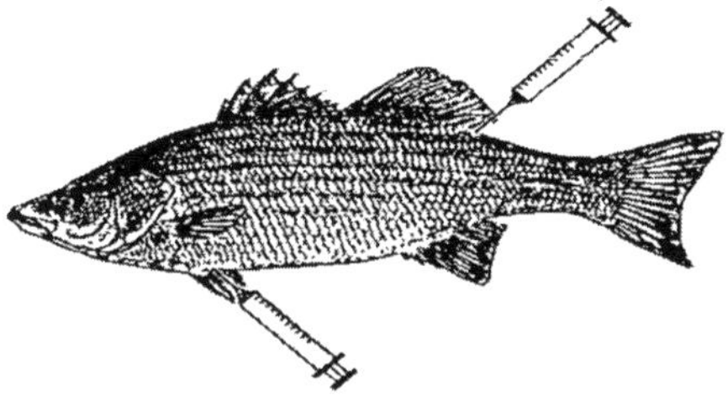

Figure 6.3. Hormone dosage and administration in producing catfish hybrids.

Source: https://www.semanticscholar.org/paper/Hormone-Preparation-%2C-Dosage-Calculation-%2C-and-for-Rottmann-Shireman/a4a20c215f05d-7c35e50385ffa58aecdf2105514

6.4 TESTICLES COLLECTION AND SPERM PREPARATION

Ill-advised taking care of and capacity of testicles/sperm can prompt poor or on the other hand zero development. As demonstrated before, it is hard to gauge testicles improvement from how it looks on the outside, especially in the blue male catfish therefore, it will require one to conduct multiple tests to determine which testicles are fit for the investigative and cross-breeding process. The male catfish is slaughtered by hitting it on the head that allows people to consume the fish without having to worry about being sedated. The worry comes in when the fish has been sedated because the sedative used might interfere with the sperm and render it useless or kill the sperms thus interfering with the process.

It is important to ensure that water does not come into contact with the sperms by drying the testicles with a clean and dry paper towel because sperms will behave differently or fail to function when they mix with water. It takes about a minute after the sperm has been introduced to the eggs to die and cease functioning. It is also important to ensure that the sperms do not die when they are introduced to new water until they fertilize the egg (Beaumont et al., 2010).

Make a mark on the male preferably on the back and create a point of entry in the middle of the body to bring forth a hole in the stomach. The testicles that are found in catfish that are matured usually appear white and are padded in appearance which those that are not yet matured have caramel pink testicles which are not huge in appearance. To remove the testicles one can either use a blade that is sharp, a tool that is considered to be surgical or by gently removing them with one's fingers. The fish will not be awake during the process and after it is done use a clean paper towel to dry the blood.

The testicles should be washed by dipping it in HBSS, using one's fingers or tweezers to remove the blood that might have clotted when removing the testicles. After cleaning the testicles, one should move them to a place where there are approximately two hundred and fifty milliliters of HBSS. Spot the testicles in a metal strainer press the testicles against the sifter with your thumb and fingers, this pounds the testicles and deliveries the sperm into the HBSS. Keep smashing the testicles until they are level and the vast majority of the sperm have been delivered. Save the arrangement, it contains the sperm.

It is advisable for people to check the mortality rate of the sperms they are going to use during the process so as to ensure that they only use sperms that are of high quality which will produce the quality results and in this case a catfish hybrid. The easiest way of ensuring that the sperms are of good quality is to use a magnifying lens. The type of magnifying lens to be used is not expensive which is an advantage because the cost of reproducing the hybrid is reduced. The lens can even be acquired from labs in schools or places where animals are treated. The process is also easy because the person is expected to pour the sperm on the magnifying lens and then use a higher resolution power to observe the sperms. However, the process should be gradual to avoid crushing the specimen. Under higher force, the sperm ought to show up as little oval \-formed specks and it is observed as a tail when one uses the magnifying lens. In normal situations, the sperms are not supposed to move and a liquid is formed. This process is known as the Brownian development. The liquid is no cause for alarm because it forms during the observation process. Do not pay attention to the movement made by the sperms. This process should always be conducted before using the sperms on the eggs because weak and low-quality sperms will affect the offspring that is produced (Beaumont et al., 2010).

When making the observation one will note that some sperms are motile while others are not, which is completely normal. That should be the quality that one looks for when trying to identify which sperms are of high quality and which ones are not. This observation happens very fast therefore it is important for the person to be very vigilant when conducting the test. When one does not observe the set expectation, they should conduct the investigation using a clean slide and it should be done quickly. This will ensure that the specimen used the second time has not been tampered with and the results are accurate to avoid mix up. Checking the HBSS is important and one should ensure that is in the right amount. In spite of the fact that motility may shift some from male to male and decreases some during stockpiling, in the event that you have no motility it is for the most part a pointer that your HBSS was inaccurately made and the sperms were activated during assortment and readiness.

Getting huge quantities of blue catfish males can be tricky since presently few have huge supplies of blue catfish accessible, yet a few makers are beginning to develop blue catfish so accessibility ought to turn out to be less of an issue soon. Little data is accessible on the presentation of cross-breeds delivered by various species of both female and male catfish. Identifying the male blue catfish has proven to be quite a task. Are the sperms obtained from these male catfish viable for making offspring that are hybrids? There is a possibility that these sperms do indeed produce offspring but only when they are carefully selected before being introduced to the eggs. It isn't known whether the exhibition of cross-breeds created from wild-caught blue males would be in the same class as mixtures delivered from tamed blue catfish, yet it is a fascinating inquiry (Beamish& Rothschild, 2009).

6.5 TIMING OVULATION

Timing ovulation is fundamental to productive blend creation. During the ovulation process, the sperms are added to the eggs for fertilization to take place. This procedure only occurs during ovulation to increase the chances of reproduction. Therefore, the best approach for a person to take in order to ensure that the process is successful, they must ensure that the female catfish is ovulating before continuing with the procedure. At the off peril that you venture to strip the female past ovulation, there will be no fertilization that takes place and the pregnancy results will be negative, on the off danger which you remain with the guide of till too long after ovulation, the egg yield is most likely appropriate anyway the high-caliber might be awful and the healing process could be low. Similarly, catfish don't release their eggs on a regular basis like other fish, which implies that catfish's ovulation process is not something that happens unexpectedly. Thus, it becomes to identify a female catfish that is ovulating and one that is ready for its eggs to be removed and used to produce hybrids.

6.6 PRODUCTION CHARACTERISTICS OF THE HYBRIDS

Harvest ability

Business venture discernments show that basically the entirety of the combos in a lake might be stuck in some seine pulls, however as a base five seine takes may not gather the entire thing off the catfish in a lake. Seinability is a belittled quality. Gain costs increase with each seining required. The gifted ejection of fish increases the reach that can be sold, facilitates the trouble of leaving fish that expands than the charming planning size, improves feed change productiveness, and diminishes setbacks.

Endurance and sickness obstruction

Numerous contamination troubles within the channel catfish industry can be decreased by using refined C×B binary when you consider that combos undergo higher in all improvement degrees. In 30 years of research at Auburn college, fingerling staying power has located the middle value of 80 to 85 percent and meals-fish endurance has arrived at the midpoint of ninety%. On-ranch persistence fees are similar. Curiously, the normal endurance of channel catfish fingerlings is 50 to 60%, and the persistence of meals-sized channel catfish is kind of 70%. regardless of the truth that the C×B crossover isn't honestly impervious to illness, it's far more safe than channel catfish to columnaris (*Flexibacter columnare*), enteric septicemia of catfish (*Edwardsiella ictaluri*), aeromonas (*Aeromonas hydrophila*), Ich (*Ichthyophthirius multifilis*), additionally, channel catfish infection. Assays show that the species might be more defenseless towards *E. tarda*, but they react directly to treatment.

6.7 SPAWNING AS A METHOD OF PRODUCING CATFISH HYBRIDS

Open-pond spawning

It is important to note that this method is not necessarily the best method of producing catfish hybrids. This is because in most cases spawning does not occur and only thirty-three percent of female catfish are spawned. Apparently, the grown-ups were not sexed appropriately when there is a case of spawning (for example, only female and male catfish were packed) and either means produced with female or males produced with blues. No mixtures were really delivered. Accordingly, this method of spawning brings forth a breed of catfish with the male catfish to create C x B cross breeds that can't be suggested.

Pen spawning

This is a more noteworthy dependable procedure of turning in CxB move breeds. Pens must resemble the ones applied in traditional channel catfish delivering. Creating pens are developed near the monetary establishment of the pond, utilizing treated wood with the pond base and plastic-covered twine network for perspectives. Male blue catfish and female channel catfish are each joined in pens while the water temperature at the significance of the delivering compartment is between 75 besides, 82o F at dawn. Male blue catfish should be put in the pens an evening before the female channel catfish. Female channel catfish picked with the model signs of status for production, for example, a sensitive, extended gut and, ideally, a genital opening that is purple and swollen. When mating happens, delivers will normally be resolved 72 hours after the female is brought into the pen. In strange cases, producing has happened insofar as seven days after the female is brought into the pen. The output has been one hundred%, however commonly achieves zero to 20 rate achievement.

Artificial spawning.

The spawning cycle can be legitimized, on the grounds that the eggs are at various formative stages which is one of a kind to each female. In like manner, the female catfish goes through the ovulation cycle. Note that this cycle possibly happens when the implantation made are pituitary mixtures. The pituitary implantations would not allow time for all of their eggs to fittingly grow first. In the creating season, the measure of LHRHa that is utilized is just 90 µg/kg and it is just utilized in the early pieces of the period. At the point when the specialists need to get the correct data during the late parts of the spawning season, the measure of LHRHa that is utilized is 75 µg/kg. Additions are implanted to the muscle which can be put intramuscularly around an inch underneath and barely at the back and at the base of the dorsal edge. Note that ovulation begins to happen at around 44 hours when implantation has been finished. Notwithstanding, the temperature of the water influences the time ovulation happens. For instance, when the water is between 75 and 77 degrees Fahrenheit, ovulation will start to occur following 50 to 55 hours. After the principal week or two of the delivering season, a couple of females will begin ovulating before the timetable as 36 hours after implantation, Slow ovulating females may regardless be giving eggs as late as 72 hours after implantation, anyway most by a wide margin will have ovulated inside 60 hours (Caliskan et al., 2015).

Before the stripping process is done to the female catfish, the male catfish which is considered to be blue ought to be sterilized. They are normally killed in a humane manner and carefully opened which is done by cutting along the butt-centric opening to a quarter of the path to the head along the tummy. This process is done with clear precisions so as to avoid cutting and harming the gut in the process. For the testicles to be eliminated, they are completely detached from the mesentery connective tissue. Eliminate just the white segment of the testicles. Dry the testicles in a tender manner, it is advised that a person uses a paper towel to ensure that the testicles do not have any blood on them and to ensure that they are not damp.

The sperm should be pressed from the testicles mixed with the female eggs at a higher speed but should be closely monitored. When in doubt, it is important to note that one male can fertilize at most five females, contingent upon the size and how the male is programmed to function. Water that has been properly oxygenated is added to the egg which are then allowed to "solidify" for almost an hour with delicate unsettling. Delicate unsettling permits the most extreme contact among the eggs produced by the females and sperm produced by male catfish. The amount of water to be added to the eggs should be monitored closely so that it does not interfere with the reproduction process. At that point, these eggs are put in a conventional incubating box.

Clearly, there is more cost associated with delivering CxB crossover catfish than channel catfish (delivered by customary lake generating techniques) In restricted field studies, the CxB cross-breed, even with the expanded incubation center expenses, could increase ranch productivity by more than 10%. The major obstructions of fast development are the absence of brood stock that belongs to the blue catfish, because basically, the entirety of the fish business depends on it, and the requirement for improving the producing strategies (Caliskan et al., 2015).

How to reduce the inefficiency in the production of hybrids?

Identifying the problems that hinder production of effective hybrids is important especially when it comes to catfish production of hybrids. There are certain conditions that increase cost of producing hybrids which is a disadvantage to many. It hinders the growth and production of hybrids that can be able to withstand harsh climatic conditions and survive in areas where the oxygen concentration is relatively low. One way to ensure that this problem is overcome is to study the different minerals found in water and their effect. Water hardness which is the accumulation of calcium in water improves the hatching process of eggs. According to the research made, when the levels of calcium are low the hatching process becomes rare and the fish become prone to diseases because they lack some important minerals to ensure that they fight diseases with ease. The levels of calcium to be used so as to ensure that hatching takes place is seventy-five milligrams in every liter.

It is also important to note that not all fish will produce eggs that are similar and even when a single fish is used the eggs will not be the same. Fish that have been given the LHRHa hormones produce eggs that are different from those that use the CCP hormones. Those eggs that do not have good genetics will be prone to attacks by fungi that will result in weaker hybrids. This will also affect other eggs that were healthy when being produced (Hart & Reynolds, 2008).

What are the genetic tools used for breeding?

Many researchers and scientists have come up with genetic maps which identify the genome found in catfish which aim at finding regions that are considered to be chromosomal. These regions are specifically designed to control those traits that are significant in these hybrids. Such traits include, the ability to produce meat and the ability of the body to resist diseases. These traits are important when choosing the specific fish that will be used.

There are DNA markers that are developed by biologists such as Geoff Waldbieser. These markers aim at identifying the parents of the hybrids and to identify diversity in terms of genetics and find those markers which are unique to certain traits. These markers help people to learn about the genetic composition of various fish which will be used to produce hybrids. Once the traits have been identified they are separated based on those that are desirable and those that are undesirable. There are other tools used which aim at ensuring that better hybrids are produced (Hart & Reynolds, 2008).

CHAPTER 7

TRANSGENIC FISH RESEARCH

CONTENTS

7.1 INTRODUCTION

The year 1974 is marked by scientists as the year that the earth began genetic innovation. With Annie Chang and Stanley Cohen at the helm, the world witnessed the first-ever transgenic organism, *Escherichia coli*. This was a result of a genetic transfer of the *Staphylococcus aureus* genes into *Escherichia coli*. But what is transgenesis? Transgenesis is a word derived from two terms, trans and genes. Trans is used to describing the aspect of moving a sample matter from one position to/into another. On the other hand, genes is a biological term that refers to DNA. The gene is the basic and most functional unit of the animal body. From this, it can be derived that transgenesis is the movement or transfer of genes from one animal to another of the same type in this case, the fish. Organisms that undergo the same, being artificially introduced with deoxyribonucleic acid and successfully introduced into their genomes are known as transgenics (Bozkurt, 2018).

Fishing dates back to history, exactly 500,000 years ago. The initial human being, the homo-erectus were the first men to go in pursuit of fish. With the understanding they had, it was almost impossible for fishing to develop not until the upper Paleolithic period. This occurred almost between 40000 and 10000 BCE years. Fish were and are still considered a major source of food. Although the earlier men were noted to eat fish raw, the future came to develop fire and learnt to cook the famous delicacy. As time passed on, fish came to be known as a major source of protein, a better form in the fact that provided essential nutrients and even medical extracts.

Figure 7.1: Fishing in the old days. Fish were plenty in the seas while the demand used to be high. Environmental pollution was not a subject of the day.

Source: Landangler's blog

The development of steam boats in the 19th century created an avenue for deep sea fishing. It was at this age that man came to realize the ability of the fish to serve in recreation. Man began to think of harvesting fish for recreational purposes. Over the ages, aquariums were developed and fish was made a source of income. Scientists came to learn of the different types of fish in the waters, noting down all the rare and common species. As his nature depicts, man became more interested in the fish that he found it interesting to develop other species of fish. But the question was, how could man make this possible?

In recent biological results, man had learnt through anatomy that genes contributed to the basic and most functional part of the ordinary animal. Creating any living organism desired that an individual comes up with genes that no one knew how to develop. It was impossible to create a gene from nowhere but it was possible to make a gene from removing and pasting parts of genes from different organisms together. In 1973, man managed to come up with a lab-designed gene. This marked the beginning of transgenesis in fish. However, transgenesis had far-reaching significance and objectives:

- Transgenesis of fish aimed at introducing a new trait of the fish which was not to occur naturally in the initial fish species. With this, man would brag about coming up with a new species of fish.
- Since the world was in a phase of evolution, transgenesis of the fish also worked in creating a media for scientific research. In a quest to develop other forms of animals, the fish would act as a perfect example as it had features that applied in many other mammalian classes.
- Because of the increased demand for fish and fish related products, the transgenic fish would also act as a means of improving food security and boosting aquaculture food production.

At this age, transgenesis on the fish was aimed at fulfilling these tasks. This brought us to 1985 when the first-ever transgenic fish were produced. China served as the pioneer of transgenesis. The Asian superpower worked hard to produce the first-ever transgenic fish. Being located in a fish-rich continent with a high demand for fish, Chinese scientists aimed at boosting food production through the Atlantic salmon, the tilapia and the common carp. Over the years, the demand grew worldwide hitting the supplied edge. The US Department of commerce and the United States national oceanic and atmospheric administration has it that there is a demand for 150million metric tons of fish (Bozkurt, 2018).

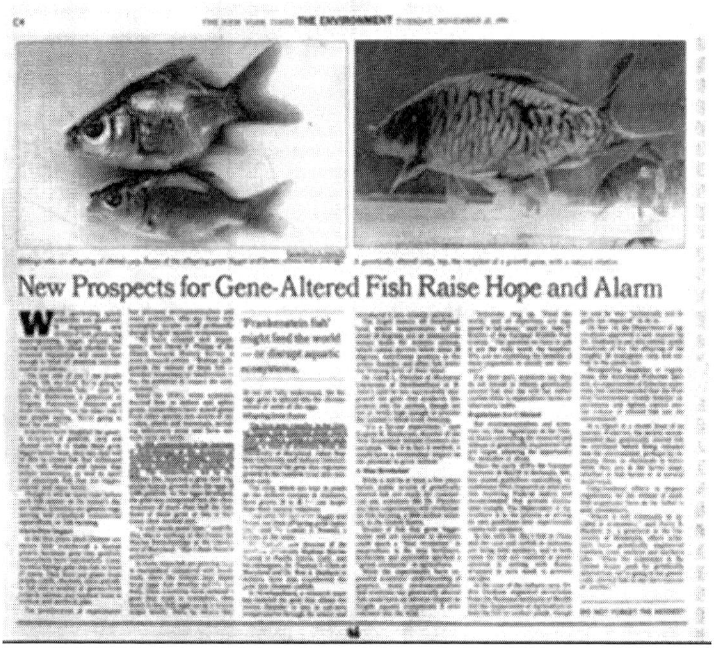
THE ENVIRONMENT

New Prospects for Gene-Altered Fish Raise Hope and Alarm

'Frankenstein fish' might feed the world — or disrupt aquatic ecosystems.

Figure 7.2: A copy of the Nov 27th, 1990 New York times which gave hope for the first-ever transgenic fish species.

Source: Institute of Hydrobiology

With time, transgenesis found its way into our hearts. This was made a success through aquaculture. Man began rearing fish in fishponds giving them the necessary conditions for their survival. It is approximated that by 1990, aquaculture had a production percentage of 7.3%, a figure that was climbing up the ladder with time. For there to be a success in aquaculture, there ought to be the introduction of transgenesis which would demand the following aspects;

- The introduction and application of modern and innovative management techniques.
- Proper disease management, i.e. early detection and efficient prevention of various types of diseases in fish.
- Revisiting and taking control of the reproductive cycle of the different types of fish.
- Introduction of clean water providing a pond with excellent and high-quality water.

- Perfect genetic background of the fish or the fish to be used in transgenesis
- Proper understanding of the fish's environment including its physiological and nutritional conditions.

If put into place, these factors would boost aquaculture to the next level. One major challenge was the introduction of modernity into the field in a world that was on the verge of innovation. If the above factors were to be considered, then operation transgenesis was a green light. Trangenesis in fish is carried out in different phases. It requires that the fish is initially identified, researched on the basis of its genetic structure and later on receiving or releasing the said set of genes. It further requires that there be means of ensuring that the fish is closely monitored to prevent any negative eventuality.

Based on recent events, it is a major factor that security be considered in any biological research. Over the past months, the small Chinese town of Wuhan has largely been under scrutiny following allegations that the dangerous covid 19 virus originated from a bat that had undergone genetic dialysis and research in a biological lab. Such cases ensure that fish transgenesis is put up to the recommended human security as fish can transmit infections 30 times the way bats can (Fletcher & Rise, 2012).

7.2 TYPES OF TRANSGENIC FISH

There are several types of genetically modified fish and the lasses that are mostly used include; Agnatha which are the jawless fish, Chondrichthyes the cartilaginous fish and the Osteichthyes, which are the bony fish.

The salmon belongs to the Salmonidae family and they include both the salmon and trout. They can range in sizes from the smallest which can be 13 centimeters i.e. 5.1 inches and the largest can be 2 meters i.e. 6.6 feet when they become fully grown adults. These types of fish all spawn in freshwater but spend all their maturity at sea. They are predators as they are known to feed on smaller fish, aquatic insects and crustaceans. An example of a genetically modified salmon the AquAdvantage salmon that was modified from the Atlantic salmon has an increased growth i.e. it matures faster and has a larger body size as compared to its wild type (Ray, 2018).

Figure 7.3: The AquAdvantage salmon. It has a bigger size as compared to the Atlantic salmon. It also matures faster.

Source: Earth Island Institute.

The modified salmon has double the weight of the Atlantic salmon and matures in half the time. As of December 12th, 2012 the AquAdvantage salmon had not been approved by the Food and Drug Agency (FDA) in the United States. The other type of salmon, the Coho salmon has also been genetically modified but it is not yet under commercial development. As of 2017, the AquAdvantage salmon has been sold in restaurants in Canada. It became the first genetically modified animal food on the market. It was proven that the genetically modified salmon has the same nutritional profile as the Atlantic salmon and was seen to be safe. It also has the same taste, texture and even flavor. Some people and scientists had safety concerns on if the genetically modified salmon would escape into the wild would it be able to crossbreed in the wild with the native wild salmon. This would never happen as the modified salmon have three sets of chromosomes (triploid) as compared to the natural one, which is diploid with the normal double sets. The triploid sets are sterile and this means that they cannot reproduce (Ray, 2018).

Figure 7.4: The ocean pout. Some of its genes have been used to improve the Atlantic salmon to help it grow more rapidly and mature faster.

Source: health benefits times.

The ocean pout *Zoarces americanus* is found in the northwest coast of the Atlantic Ocean. It is known to have antifreeze protein in its blood which gives it the ability to live in near-freezing waters. Scientists have taken the genes from the ocean pout and implanted them into the Atlantic salmon. The anti-freeze protein gene is used in conjunction with a growth hormone found in Chinook salmon, which then makes the blood concentration of growth hormone to be of higher concentration in the blood. This then leads to this genetically modified salmon to grow more rapidly.

Tilapia is a species of cichlidae fish family. They are found in freshwater habitats which include lakes, streams, rivers and ponds. They live in warm temperatures and are found in warm temperate regions, they are vulnerable if exposed to cold temperatures, and they survive well in temperatures above 60^0 F (16^0C). They are the fifth most important fish species with production averaging about 1.5 million metric tons in the early 2000s. This is because they are large in size, they grow relatively fast and are palatable. They have a mild-flavor and are lean. As of 2016, it was the fourth on the list of seafood that people in the United States were more likely to consume. They are more preferred because they are easily adaptable and they can even survive in water that is of poor quality and sometimes in overcrowded conditions. The most popular type of tilapia that is farmed is native to the river Nile in northeast Africa. They are mostly raised in enclosed tans to prevent them from escaping. Raising them in cramped and overcrowded conditions can lead to diseases. They feed on algae but can also survive on several varieties of food (Ray, 2018).

Figure 7.5: Genetically Improved Farming Tilapia. They are fast maturing, have increased body size and more resistant to diseases.

Source: tilapia market

There are several sources of tilapia in the world and they include; the tilapia Recirculation Aquaculture Systems (RAS) farms in the united states, blue tilapia raceways in Peru, Nile tilapia open-net pen farms in Indonesia, Mexico and Honduras and tilapia pond farms in Ecuador. Worldfish and their partners created an improved strain of tilapia called Genetically Improved Farmed Tilapia commonly referred to as (GIFT). The first generation of GIFT was bred to grow fast and reach harvesting size quicker. The research on the improvement of GIFT was based on selective breeding. This method of selective breeding through the GIFT project is known as the GIFT technology and has been applied to Nile tilapia and other species in Egypt, Malawi and Ghana. There are intentions to start improving the GIFT fish using a DNA-based approach and genomic tools to improve specific aspects of the fish. When DNA is used to alter the fish characteristic the tilapia has been seen to have no abnormalities. Tilapia is also an excellent source of omega-3 fatty acids and proteins.

The zebrafish is also one of the genetically modified fish. It is a freshwater fish and is part of the family Cyprinidae. They are popular in aquariums. Their name is derived from the horizontal stripes along the sides of the body's bilateral and they are also uniform. The males and females both have stripes but are of different colors with the males having gold stripes within the blue stripes and females having silver stripes along the blue stripes (Ray, 2018).

Figure 7.6: The original zebrafish. A male has gold stripes with blue stripes that are horizontal.

Source; University of Stirling.

They can grow up to 6.4 centimeters when mature in the wild but can only grow to 4 inches when raised in captivity. There are two main types of zebrafish that have been modified *Brachydanio rerio* (GM ZF) and medaka (*Oryzias latipes*) which have the fluorescent gene introduced to them. They are commonly referred to as GloFish and it is a patented trademark of genetically engineered fluorescent fish. The right is owned by Spectrum Brands Inc. these fish are now sold in many different colors such as bright red, green, and orange-yellow, blue, pink and purple fluorescent colors. The source of the fluorescent protein include, *Aequorea Victoria* (a jellyfish), *Renilla reniformis* (sea pansy), *Discosoma* (mushroom coral) and many other organisms. Studies have found that wild males had an advantage over the modified fish as they were found to be more aggressive. The wild males sired twice as much as the modified fish. The females on the other hand preferred the glofish as compared to the wild zebrafish. In terms of predation the bright red modified glofish were twice as vulnerable as compared to the others. This type of fish was first intended to be used as a way of detecting pollution. Although it is illegal to reproduce these fish it was stated that the eggs of the fluorescent fish are pressure treated to make them infertile but it has been noted that glofish are indeed fertile and can reproduce in a captive environment (Bert, 2007).

Figure 7.7: The different colors of modified zebrafish. They are mainly ornamental but they were meant for the detection of pollutant toxins found in water.

Source: Ardea

7.3 TRANSGENE CONSTRUCTS.

A transgene used in the production of transgenic fish is always taken into high consideration. It ought to meet the threshold of being a recombinant gene construct that eventually produces a gene product that will fit into the desired set of tissues and achieve compatibility at the desired time. This is made possible by the introduction of an appropriate gene promoter enhancer which is contained in a plasmid and contains a structural gene sequence.

There are three main types of transgenes;

- *Reporter function*
- *Gain-of-function* and
- *Loss of function*

Just as the name suggests, the reporter functions are used for the identification purpose. They give the scientist information telling him/her of the success of the gene transfer process. It is also used in the effort of measuring the strength of a gene (promoter enhancer element). To sum it up, the reporter function transgene awards data determining the transcriptional regulatory sequence of a said gene. The gain-off function transgenes just as the name suggests work to add new functions to the transgenetic fish and facilitate the identification of the said fish. For instance, if the genes of a fish that tend to swim upright or upstream are removed and transferred onto another fish, the transgenetic fish will acquire the new trait of swimming upright. This explains that the fish would have obtained other new functions.

The loss of function transgenes interferes with the operation and expression of the host genes. The loss of functions transgenes have the tendency of interfering and encoding the ribonucleic acid and in turn interfere with the post-transcriptional process that works to translate the endogenous RNA. They can also encode a catalytic RNA in a host fish referred to as a ribozyme that tends to cleave specific sets of fish mRNAs. This process will in turn cancel the production of the normal gene product limiting the fish of its initial functions hence the name loss of function (Bert, 2007). The loss of function genes has never been applied in fish transgenesis but the same is expected to apply in the future. The loss of function transgenes will work a long way in reducing fish death due to diseases as it will create disease-resistant transgenic broodstocks. With the improvement of aquaculture in the modern world, farmers will be encouraged to apply the same to limit disease transmission and unnecessary fish illnesses. The loss of function transgenes is also expected to create a transgenic model of fish that would be defective in certain kinds of gene products. Such sets of fish would be important to the research department in any institution (Bert, 2007).

7.4 DETECTION OF TRANSGENES

There are various ways in which a scientist can be able to detect transgenes in fish. Those ways include northern blot, southern blot in what is known as *situ* hybridization, slot expressions essays and polymerase chain reaction.

Most scientists that have ventured into understanding and detecting transgenes have mostly used the hybridized slot which involves the southern blot and the northern blot. Of the two, the most common one is the use of the southern blot hybridization. By use of the southern blot hybridization, involves generating the DNA fragments by a process of digestion restriction that are then added to a gel known as agarose in which the electrophoresis of the gel is used to help in the detection of the introduced gene. After the introduction of the same in the agarose gel electrophoresis, the fragments of the DNA get separated and then transferred to a membrane that is nylon-like by a technique known as blotting. The membrane can then be hybridized by a radioactive probe which will help the scientist see or rather be able to find the DNA of interest. In such situations, the radioactive probe that is used in hybridizing the membrane is normally that which has similar homology as that of the membrane. Apart from the fact that the radioactive probe can be used in the detection of rearranged genes in a certain DNA, then can also be used in the identification of genes that are related structurally in a singular species of animals that are similar or homologous genes in different species of animals. This can be seen in a situation in which there is DNA fingerprinting (Beaumont &Hoare, 2008).

With the case of the northern blot, the mechanism is almost similar to that of the southern blot. However, in this case, instead of using DNA fragments, scientists use RNA to determine the genes of the fish. In this case, the radioactive probe has the ability to measure RNA transcript accumulation that is considered to be helpful to gene expression studies. Slot and southern blotting are also quite similar. As seen earlier, southern blotting normally requires a genomic DNA cut with the restriction enzymes before they are moved to the nylon membrane. This brings the difference between southern and slot blotting as slot blotting does not require the DNA to be cut prior to its transfer to the nitrocellulose filter or the nylon membrane. This makes it easier for scientists to do the analysis for both the multiple samples and in other cases the DNAs that are degraded. In most cases, scientists tend to prefer southern blot hybridization to slot blotting because of the fact that slot does not provide a lot of information. As a matter of fact integrations or rather rearrangements of the transgene will not be indicated when one is using slot blotting hybridization.

There is another type of blotting known as western blotting. Its function differs from that of the above three types of blotting. As seen, the above three types of blotting are relevant when detecting the transgenes in the DNA of a certain species of animal. For the case of western blotting, the primary function is to characterize and identify a specific gene product. The process that is used is known as electrophoresis, the process involves transferring proteins from a gel known as acrylamide to a membrane known as nitrocellulose. To detect the protein that is of interest, the nitrocellulose membrane is then probed with a certain antibody. Another way that a scientist can detect transgenes is through *in situ* hybridization. In this case, nucleic acids that are found within the cytological preparations in the body of an animal is hybridized. This method has a rather different function from the various blotting techniques as it helps scientists to identify how the transgenes are localized (Beaumont &Hoare, 2008).

The polymerase chain reaction is rather based on a totally different concept. It is based on the amplification of different fragments. Polymerase chain reaction amplification is rather based on a continuous cycle of denaturation. It is also based on continuous cycles involving the annealing of certain primers known as an oligonucleotide that is rather complimentary to a certain gene and also a primer extension that is done through Taq polymerase. Once the fragment has been amplified by polymerase chain reaction then can then be easily recognized as a discrete fragment on a gel or in other cases on a southern blot. The process of polymerase chain reaction is considered to be very interesting and further is considered to be a learning process. As such, its application by a scientist in the research involving transgenic fish continues to grow. The process is however very delicate and the specimens used in the analysis can be easily contaminated. As such, the research team is obliged to have certain strenuous measures that would help avoid the contamination of the specimen.

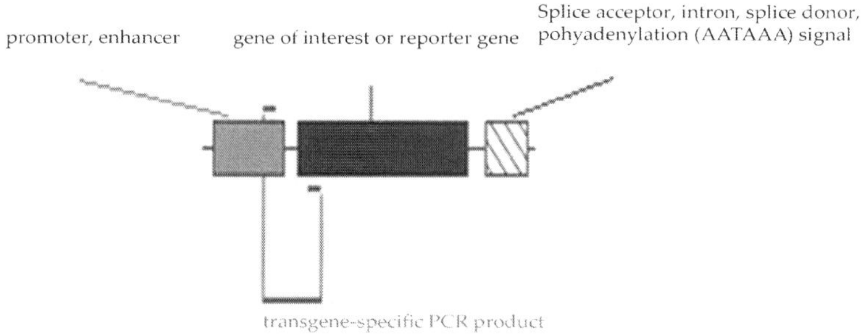

Figure 7.8: Polymerase chain reaction is one way of detecting transgenes.

Source: University of Virginia School of Medicine.

It should be noted that the polymerase chain reaction method is considered to be a more sensitive method in the screening of a transgenic founder fish as compared to southern blotting. Contrary to that, to ensure that the analysis is reliable, scientists prefer the use of southern blotting. To detect a transgene may also prove to be quite difficult or rather not straightforward. This is so because there is a transgene copy number for them to be easily detected. Another thing that may cause difficulty in detection is the fact that the transgene in most cases tends to have a certain mosaic distribution in the host fish. As such, to detect the transgenes, most scientists use the reporter genes since the gene product can be readily assayed. Scientists have extensively used reporter genes which include those that code for lacZ. The lacZ enzyme has proven to be quite reliable as it can be easily visualized in embryos that are whole to detect luciferase activity and also in the process quantifying it, scientists use a scintillation counter (Gupta, 2008).

Reporter gene products such as beta-gal and CAT are found in certain host cells and to detect their presence, scientists use Expression assays. The thin-layer chromatography method and the two-phase partition assays are the two methods that most researchers use in the detection of CAT. Two-phase partition is dependent on the diffusion of the chloramphenicol acetyl that is labeled with a liquid that is water-immiscible by a technique known as scintillation. As an acetyl donor, the two-phase partition assay utilizes [^{3}H] or rather [^{14}C] acetyl CoA. The process is considered to be quite simple and also convenient to use but it is clear that it has not been utilized widely in the research involving transgenic fish. The most popular method in transgenic research is rather the thin-layer chromatography method used in measuring CAT. Fluorometric determination techniques and in situ X-gal staining are some of the techniques under thin-layer chromatography that are used by scientists in detecting the presence of beta-gal in fish cells of different species of fish.

To detect the product of the transgene expressing, which in this case is the beta-gal, in fish cells scientists use the X-gal staining technique. If the beta-gal is present, the stain in the X-gal is cleaved to form deposits that are localized and are of indigo dye. On the other hand, fluorometric determination is quite important as it is used to measure the hydrolysis of 4-methylumbelliferone-beta-D-galactoside by use of beta-gal. The methylumbelliferone is connected to a certain fluorogen and thus can be easily measured quantitatively by use of a fluorometer. What is known as the GFP gene is considered to be quite important in transgenic fish research. This is so because the protein that is present in this gene can be detected in live organisms by inducing the protein to fluoresce. Another important detail that should be noted is that the presence of a functional promoter can be indicated by the expression of the reporter genes. When one decides to establish the expression of a growth hormone, it can be done through the reverse transcription of the polymerase chain reaction method which helps in the detection of the mRNA. It is followed by the hybridization of molecules by the use of a gene-specific gene hormone by the process known as radioimmunoassay. The passing of a transgene onto a progeny needs to be established by the process in which genetic crosses are done between a wild type of fish and a founder fish that is carrying a transgene and is considered to be the mate of the other. This is what is known as germline transmission. It shows that through mating, there can be the transfer of transgenes to their progeny. This will be discussed further in the later sub-topic. To prove the transfer, the progeny of the two fish are further tested to see if there are transgenes in their cells and also tested for their expression (Gupta, 2008).

7.5 METHODS OF GENE TRANSFER

There are various methods in which genes can be transferred and ranges from internal ways to external ways. As a matter of fact, scientists have come up with various methods that have been used to transfer genes into fish eggs. The method of gene transfer can also range from a sing fish, few fish to a mass gene transfer. Those methods include the disruption of the egg membrane by use of microinjection, electroporation that involves a pore formation in the cell membrane in which it is done through exposure to a high voltage, puncturing of multiple membranes by use of charge interaction between the lipid and the membrane or through micro projectiles. Other methods involve the use of male fish that can then transfer the genes through the process of fertilization. The most common method used in transgenic fish research and rarer is the use of microinjections.

Microinjection of single cells

This method involves transferring the DNA to an individual embryo or eggs. The injection that is administered normally depends on the size of the egg. However, the standard amount injected is normally200pl to 20nl. Also, there are a number of copies used in terms of the amount of DNA injected but it rather normally varies with the species of fish eggs and the size of the eggs. In most cases, the DNA amount injected may be up to 10^6. For fish, the injection is different as compared to other animals. For instance, for the case of the mouse (involving their embryos) the injection of the solution containing the DNA is done into the pronuclei. This is so because the pronuclei of the mouse are quite visible unlike that of fertilized fish eggs. Furthermore, for most fish species, the eggs tend to be opaque. It is because of this reason that the solution containing the DNA is injected into the cytoplasm. In some fish species, microinjection of the fertilized fish eggs may prove to be quite difficult. This is so because some of the fish eggs have a hard chorion which makes the process quite difficult.

Scientists have however come up with various measures that can be useful in curbing such problems. The injection may be done through another membrane whereby the injection is subjected to the embryo through the microphyl. Another method involves the crushing of the hard chorion by the use of a hard needle before the insertion of the glass needle containing the DNA. There are some scientists that have gone forth to manually eject the hard chorion from the fish eggs or digest the chorion through the use of pronase before injection of the DNA strand. After the injection of the DNA into the fish embryo, the survival rate tends to be different with different species of fish. For instance, in zebrafish, the rate of survival is 16% while in salmon, it is 85%: the production of transgenic founder fish is normally done in a laboratory. The number normally depends on the laboratory and the species of fish that are used. For instance, if they use zebrafish, the amount grows by 5%, 6% if salmon, 12% if carp, 13% if tilapia and over 70% percent if rainbow trout is used. To ensure the survival of the fish, it should be ensured that the gene transfer is efficient. There are various actors that can however affect the efficiency of the gene transfer. They include:

- The buffer used in suspending the DNA
- The concentration of DNA used
- The use of either a linear or a circular DNA

An example of such a case involves the rainbow trout species. When this species was injected with a DNA strand known as EDTA, the survival rate of the embryos of these fish radically declined but its retention of the original DNA increased from about 10% to 13%. As compared to the use of 10^6 copies, the use of 10^8 copies led to the decline of survival rates by 10%. The use of linear DNA was considered to be more efficient by 15-fold in integration as compared to circular DNA.

The microinjection method of gene transfer is considered to be a laboratory method. An experienced lab operator may for instance be allowed to inject about 100 zebrafish in a single hour. When however there are two operators, they are allowed to inject 60 zebrafish at a single moment. As such, scientists have come up with various methods to conduct mass gene transfer, but have proven to have limited success (Dunham, 2011).

Mass gene transfer

- Liposome-mediated gene transfer (lipofection)

  In this case, the negatively and the positively charged liquid DNA molecules interact under this method, there are three models that scientists have proposed to involve the internal model, the electrostatic model and coated electrostatic model. Under the internal model, DNA is entrapped within the liposome. It also involves electrostatics whereby the DNA interacts with the inner surface of the liposome. With the electrostatic model, it helps in prediction whereby it predicts that DNA and the cationic liposomes interact through charges that are electronic in nature.

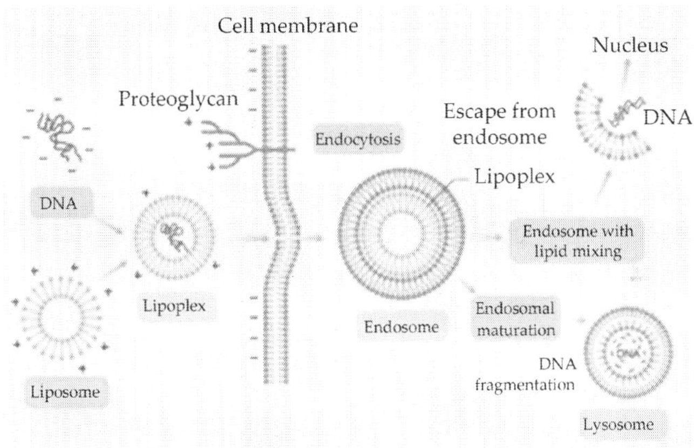

Figure 7.9: Negatively and positively charged liquid DNA molecules interact under this method.

Source: BrainKart

The third model however does not involve electrostatics. It claims that DNA is surrounded by double lipid bilayers in which the nucleic acid is enclosed in a ribbon. Surfaces of most biological membranes tend to have a negative charge. As such, the liposomes that are contained within it can easily associate with the negatively charged plasma membrane. As such, the nucleic acid within it can be easily transferred into cells. This process is however a bit confusing as most scientists claim that they cannot easily comprehend it. To determine the rate of uptake, then one needs to consider the rate at which the liposomes can bind to the plasma membrane. The specificity of lipids on certain specific cells also plays a crucial role in the process of uptake. This was done by various scientists on an African catfish. The process was however considered to be very unsuccessful as with increase in age, there was a decline of the genes in the fish. As a matter of fact, after a period of 4 months, the transgene was no longer present as they could not detect it anymore. As animals tend to degrade as they grow, it shows that all the introduced DNA strand was lost in the process (Dunham, 2011).

- Particle bombardment

  This method involves the coating of the DNA that is to be transferred to a certain fish on either gold or tungsten particles. They are then kept on a bombarded device and are then discharged to the eggs. In this process, the fish eggs should be under a vacuum. The distance between the eggs and the discharging equipment plays a major role in ensuring the efficiency of the process. It is normally done on fertilized eggs of different species of fish ranging from zebrafish, loach, brine shrimp and rainbow trout. This method is considered to be quite effective as the survival rate of the fish is quite high and further, there is the detection of the genes after the fish have hatched and grown. However, there was a problem as most of the embryos did not grow and experienced stunted growth. This process is in its starting ages and is still a work in progress.

- Electroporation

  This method involves the exposure of cells to an electronic field. This makes the components of the cell membrane become polarized. As such, there will be a development of a voltage potential which creates a difference between the inside and the outside. This will cause the breaking down of the membranes in areas that are localized which then leads to the formation of pores. These pores are however considered to be temporary as they can be reversed as long as the electronic pulse is lower than the critical limit. The pore formation by use of electronic pulse is what is known as electroporation.

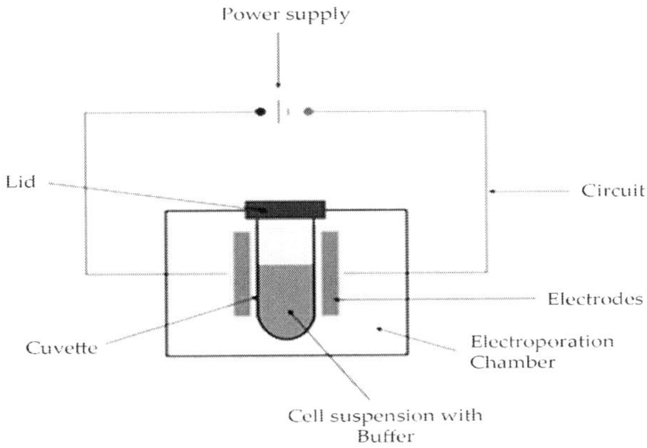

Figure 7.10: Electroporation is a modern gene transfer technique that is even used in transgenic fish research.

Source: Genetic Education

- Sperm cell as a vector

  This is one way in which mass gene transfer in fish can be done. This is easily described by the mechanism in which fish procreate. The female fish engage in the process of laying eggs and the male fish then later come to fertilize the eggs. If the male is implanted with a transgene, they can distribute the sperm to the laid eggs and the gene transferred to their progeny. Apart from fish, this method has been used in different species of animals.

7.6 IDENTIFICATION OF TRANSGENIC FISH.

They say that building a house is not that hard of a subject, the hardest part would be moving into the house. So is the case with transgenesis in fish. Identifying the kind of fish for the process, coming up with ways of transferring the genes and the process of transferring the genes are just mere processes in the entire program. Once the above is accomplished, the fish farmer or scientist is entitled to find it and identify the successfully transgenified fish. Traditionally, this could be determined in two broad ways, the dot-blot or the southern-blot hybridization of genomic deoxyribonucleic acid (DNA) obtained from the tested set of fish. This was conducted by obtaining samples of genomic DNA from the tissues of test subjects in the school of fish. The obtained samples would then be digested with restriction enzymes and the dot-blot. One of the disadvantages of this method is that it is very expensive. The effort in terms of finance is so big that scientists would tend to avoid going through the process. It also required much labor as many people would be entitled to work in the labs with very long shifts. The process was also very insensitive and discouraged many from fish transgenesis. However, this was the only way of identifying whether a fish had obtained the said genes or whether the genes had integrated into the fish (Ray, 2018).

The identification of transgenic fish aided in lab research by identifying the pattern of transgene integration. This would be the case if appropriate restriction enzymes are employed in the southern-blot analysis. In the modern world, there has been the introduction of the polymerase chain reaction (PCR). The PCR is effective in that it gives a platform for one to handle a larger number of samples as compared to the older means. The PCR is also very economical in that it does not cost as much as the dot blot and the southern-blot hybridization.

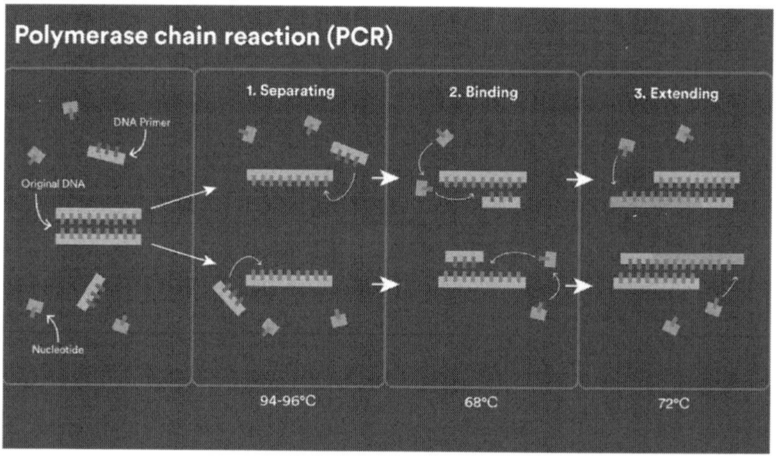

Figure 7.11: A diagram showing how a polymerase chain reaction works.

Source: Newsela

This process first requires that genomic DNA is obtained from a small piece of fin tissue of the fish in question. After the genomic DNA is obtained, there is PCR amplification of the transgene sequence before there is a southern blot analysis of the amplified set of products. This method however has some limitations. It does not differentiate whether the gene is integrated into the fish or if the genes remain as a set of extrachromosomal unit with no apparent functionality.

7.7 EXPRESSION OF TRANSGENES IN FISH

There are five ways of detecting transgene expressions;

- RNase protection assay
- RNA Northern or dot-blot hybridization
- Reverse transcription polymerase chain reaction (RT-PCR)
- Immunoblotting assay and
- Biochemical assays for determining the presence of transgene protein products.

RT-PCR is the most common means used as it requires only a small amount of sample making it cheap and efficient in most laboratories. First, a ribonucleic acid (RNA) is ejected from a small piece of tissue. A single-stranded cDNA is synthesized by reverse transcription and the PCR of the transgene cDNA amplified by applying a pair of oligonucleotide primers specific to the transgene set of products. This method should be well managed to avoid contamination of the genes in the reaction (Ray, 2018).

7.8 PATTERNS OF FISH TRANSGENE INTEGRATION

Research has it that most species of fish experience the art of transgenes maintained as extrachromosomal in relevance to the injection of circular or linear transgene constructs into the embryos of the fish. The pattern begins by their being fish tissues or embryos from the fish. The ribonucleic acid is then extracted by mincing and boiling or by any other relevant means. After extraction, the RQ DNase is treated before being applied with DNA polymerase. After that, there is a series of reverse transcription that results in the elements being added with a chelating mixture. The components can finally go through the polymerase chain reaction.

In recent studies, many fish types and species have recorded transgenic integration in a head to head, head to tail and tail to tail form. However, there were some exceptions such as in the common carp and the channel catfish.

7.9 TRANSGENES INHERITANCE.

It is the aim of most scientists and aquaculture farmers that the invented form of transgenetic fish is passed on and reproduction continues. This requires that there is a stable integration of the transgenes in a fish meeting the requirements for continuous vertical transmission to subsequent generations. This assures the fish of micro-creating a transgenic fish line which will benefit the world of research and food security in the near future. To find out if the above is a success in a fish species, the transgenic fish species is mated to an ordinary fish species. The resulting progeny is assessed and assayed to find out if the fish has the presence of transgenes by the initial transgenic fish species.

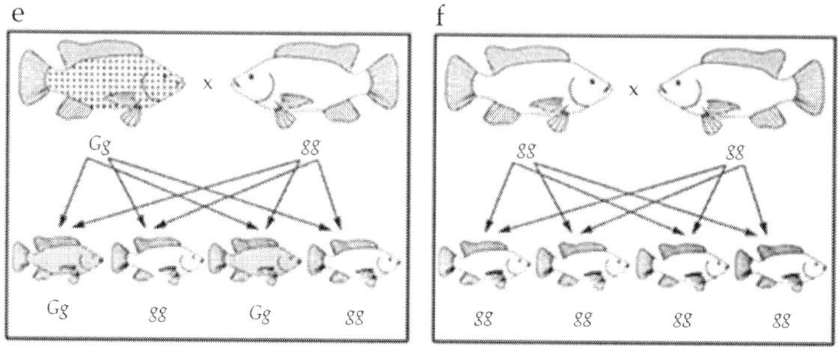

Figure 7.12: A diagram showing how genes are transferred in a fish

Source: Selective Breeding Programs

This aspect upon success is repeated in future generations of the fish to realize if a stable fish line is created. The threshold however requires that the germline of the transgenic fish gets transformed with at least one copy of the transgene per haploid genome. It is required that there be at least 50% of the initial fish's transgenic progeny. However, most tests record about 20% of progeny matches telling of the work required to fulfill the transgenic threshold. A successful inheritance suggests that the transgenic fish are generally mosaic and that the same is accomplished as a result of delayed transgene integration during phases of embryonic development (Ray, 2018).

7.10 BIOTECHNOLOGY AND THE APPLICATION OF TRANSGENIC FISH IN BIOTECHNOLOGY

In the modern world, biotechnology has captured the attention of mankind. The creation of new species and the development of new gene-related species in transgenesis is a key pillar of biotechnology. Many may argue on the importance of biotechnology in aquaculture yet it is the dependable factor for most farmers in aquaculture. There is a tendency for most animals and fish included to adapt to environmental changes and become less tolerant to diseases and other alarming factors. The fish are a known key figure in the food chain whose dependability is known in the seas and other water bodies.

If biotechnology allows for fish to perish, there would be a gap that would find a way of hurting the world going to the extent of affecting millions of the human population who are totally dependent on fish for food.

Transgenic fish helps us flow in similar motion with technology, creating fish strings that can adapt to environmental changes and resist various environmental changes. It would help preserve the fishing culture for future generations. Biotechnology has also awarded us a way of identifying both the deoxyribonucleic acid and the ribonucleic acid in the labs through electron microscopes. Away from the gene, the fish are provided with a favorable environment that ensures their survival in the research program. Taking into mind that there are other fish species on the verge of extinction, biotechnology has a way of ensuring that the fish are put into consideration awakening the future generations (Beaumont et al., 2010). In conclusion, there is a great deal of potential and capability lying behind transgenesis in fish. It is compared to a future where man gets what he/she wants. If one aspires for a recreational fish that would fit a certain desired need, then transgenesis is the solution. Transgenesis of fish is the solution to revolutionizing the aquaculture industry. In a world fitting the demand of 150 million metric tons of fish, it is almost inevitable to say that aquaculture should be practiced in every single part of the world.

Some of the great advantages of the transgenesis of fish are that transgenesis would ensure that aquaculture bears a growth enhanced generation with improved food conversion efficiency ensuring of food security in the future world. Transgenes would also work to resist current fish viruses adapting to changes in their own environment. Over the past years, the world has recorded a drop in the oxygen levels in the high seas. Being the largest provider of fish, the lack of sufficient oxygen in the high seas has been detrimental to aqua-survival. However, coming up with transgenic fish allows us to fit in genetic variance that will allow the fish to survive on the minimal amounts of oxygen available underwater. Moreover, the recent change in seawater temperature as per the change in environment can be cured by transgenesis. Transgenesis will allow the fish to survive in the coldest of places. Transgenic fish can tolerate sub-zero temperatures. Research has it that transgenic fish tend to grow faster than ordinary fish and are in fact more profitable than natural fish. It is however alarming how some labs tend to push for the destruction of other natural fish species. If we destroy nature, then it will always find a way of hitting back at us. In our quest to promote transgenesis in fish, we ought to respect the water kingdom and every natural thing it entails (Beaumont et al., 2010).

CHAPTER 8

SECURING THE FUTURE OF SEAFOOD THROUGH GENETICS AND GENOMIC APPLICATION

CONTENTS

8.1 INTRODUCTION

The fundamental concept behind genetic modification is to create a new set of DNA known as genetic "construct", which consists of a promoter or switch. It introduces the required gene into the transgenic seafood, and allows the reporter gene or marker fragment to enter the genome. The geneticist detects the existence of the new gene, including the terminator section that terminates the gene.

The real gene function is much more complicated. For example, gene products may impact other parts of the organism and produce unexpected results, the switch may also trigger or inactivate genes other than the required transgene, and also produce unexpected effects, plus the location of the construct into the seafood's DNA is also unrecognizable and may be unreliable (Zargar, 2020).

Scientists are looking for ways to genetically modify fish and other sea organisms to introduce or expand features of economic value. Fish is of particular concern to food researchers because most fish produce large numbers of eggs. These eggs are outside the animal, making it relatively easy to insert new DNA.

At present, there are around 35 sea fish species and various mollusks, plants, crustaceans, and marine microorganisms being studied globally for genetically modified strains. The sea fish are genetically engineered to enhance food production, produce medicines, test water pollution, and other unique purposes.

Genetic engineering technology allows for the manipulation of genetic characteristics to modify and develop organisms. Various genetically modified fish and seafood are currently being developed, withpossible benefits, such as increasing the productivity of aquaculture and enhancing human health. Nevertheless, some individuals worry that in the rapidly developing field of GMO, current technical and regulatory protection measures are not enough to safeguard the environment and guarantee that the public accepts these products for safe use (Zargar, 2020).

During the early 2000s, various efforts were made to genetically modify fish and seafood items, among which the genetically modified AquAdvantage salmon was developed to create new products for human usage. Environmental issues related to the creation of genetically modified fish include the probability of unfavorable competition with wild fish, and the possibility of hybridization with wild fish, allowing modified genetic content to escape into wild fish populations. Sterilization and biological control have been suggested as techniques of separating genetically modified fish to reduce the danger to wild fish populations. For instance, some scientists recommend reproducing salmon eggs (all sterile female eggs) in Canada, transferring these eggs to Panama, nurturing and processing fish in Panama, before shipping the edible processed fish to the U.S. for retail (Zargar, 2020).

8.2 RISK AND BENEFITS OF GMO SEAFOOD

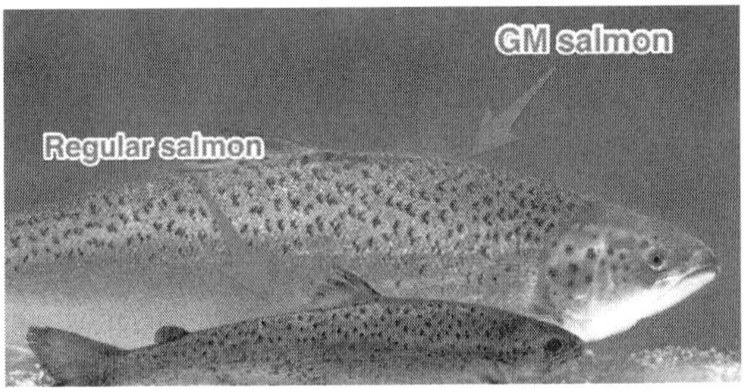

Figure8.1: Genetically modified salmon matures faster and is more commercially viable than the wild species

Source: https://www.businessinsider.com/fda-approves-gmo-salmon-2015-11?IR=T

Genetically modified seafood has a great capacity to further grow the output of fish farms, but it has aroused serious concern about the possible environmental effect of wild species in Europe and the United States. In order to overcome these issues and address the public's opposition to GMO seafood, it is therefore essential to develop a reasonable, reliable and broadly accepted technique to estimate the possible danger caused by genetically modified seafood entering the wild.

Researchers have developed a variety of genetically modified seafood, but none have been endorsed for aquaculture. For instance, the main fish mentioned by critics is Atlantic salmon, which has a growth hormone genome derived from Chinook salmon. According to its developer and producer Aqua Bounty Technologies, Inc., juvenile GMO fish grow 4-6 times quicker than wild salmon. Other genetic adaptations in various seafood species can better resist bacterial infections or make them resistant to cold conditions. Improving nutrient utilization is another significant research area. In addition, genetically modified seafood is being created for the biological control of invasive species. One means of targeting particular species is genetically transforming "Trojan horse" genomes into genetically modified seafood and releasing them, so that the genetically modified seafood will find a way to enter and influence the invasive population (Gjedrem, 2005).

Apart from commercial genetically modified seafood, transgenic seafood has been extensively developed and used in different laboratories across the globe as a model for comprehending growth and development systems, disease resistance and studying human infections. Proponents of genetically modified seafood mention that when any such genetically modified seafood is used in aquaculture, there are problems with the global depletion of seafood resources. The Food &Agriculture Organization (FAO) of the United Nations approximates that the global demand for sea fish grew to 110 million tons in 2010, mostly due to population growth in Africa, Asia, and South America. The majority of these fish would have come from aquaculture and genetic modification labs.

However, more extreme seafood farming practices may create environmental problems, mostly through runoff waste and population concentration, which may become a breeding ground for the emergence of viral illnesses and parasites (like sea lice), and may also endanger wild seafood. In order to partially solve the problem of waste loss, cold extrusion float feed has been made. These feeds are neither biodegradable nor edible, nor sink down to the bottom. They are now prevalent in the market, combining computer and video-based feeding systems to gauge feed intake within the water and close the feeding ecosystem when the pellets begin to fall below the water level of the seafood fed.

Additionally, most countries require 1-2 dives per year. During this period, the regulatory agency monitors the seafood living conditions. If sludge is produced, adaptive management is required. Likewise, the use of antibiotics has been substituted by vaccines. In the 1990s, the use of antibiotics in Norwegian aquaculture fell by 98%.

Proponents believe that genetically modified seafood can further alleviate some of these concerns by offering better disease resistance, quicker growth and better food use. Obviously, some of the modifications are mainly to increase the economics of seafood farms. Using seafood that grows faster can enable farmers to produce more seafood every year at a lower cost. Likewise, cold tolerance allows farmers to diversify aquaculture to colder and sparsely populated regions, like northern Norway or northern Canada, however it does not essentially improve the ecological impact of seafood aquaculture.

In this regard, transgene seafood is more promising. Disease resistance will enable farmers to further reduce the use of antibiotics, insecticides and fungicides. Besides, improved nutrient utilization will reduce the impact of seafood aquaculture on the nearby environment. In addition, the purpose of enhancing nutrient utilization is for exploiting previously indigestible compounds, like phytic acid as a possible source of phosphorus, for reducing pollution and the cost incurred by seafood farmers. However, no matter the kind of genetic modification applied, the key benefit is still its impact on native species-so long as the farmed seafood can be prevented from escaping-any development in aquaculture will reduce the pressure on marine fisheries (Gjedrem, 2005).

Opponents admit this argument, but still believe that genetically modified seafood poses a great risk to wildlife. If genetically modified seafood escapes from the farm, they may further damage the fragile ecology of the ocean, leading to ecological destruction or species loss. A genetic modification that increases cold, salt and heat tolerance can expand GM seafood into new areas. Genetically modified seafood has greater disease resistance and improved nutrient utilization, they can outdo their wild relatives and interfere with the predator-prey relationship. These GM seafoods can occupy new niches where wild species typically cannot survive. Lastly, by reproducing with wild species, the escaped genetically modified seafood may disperse the transgene across the wild population, which may have varying impacts on mating success, the viability of the natural habitat, and other adaptive factors needed for the species survival.

In view of this, the escape of non-GMO domesticated seafood may create as much peril as the release of genetically modified seafood. Breeding of domesticated seafood and choosing artificially raised seafood has been done to keep predators away. When the seafood is abundant, they can easily rebound from depletion and avoid wild predators from reducing their numbers. However, if these animals are mixed with their wild counterparts, genetic crossing may occur that taints the wild gene selection and causes a general decline in the adaptability of the whole population. In the worst circumstance (which may have occurred in some places), the gene bank becomes so contaminated that "natural" populations must rely on commercial distribution to be viable. This contrasts with well-managed hatcheries that avoid artificial selection to prevent this problem.

However, the main reason for a hatchery-based supplement or restoration plan is to reduce habitat loss. Next, it helps offsetting decades of over-harvesting by introducing more seafood and allows commercial producers to continue working. Obviously, caution is needed. Once genetically modified seafood escapes into the high seas, it is obviously more difficult to control and spreads faster than terrestrial genetically modified plants. Even if genetically modified seafood is kept only in a safe pen, it cannot be disregarded that human error and natural calamities (such as storms) may cause them to escape. But this concern could be exaggerated, since escaping seafood usually starve to death before learning to find natural prey instead of feed pellets. Therefore, the first step to assess whether the possible environmental risks of genetically modified seafood will exceed the potential benefits is to create a reliable and practical risk assessment method (Gjedrem, 2005).

In fact, it is easier said than done to determine whether the escaped genetically modified seafood will cause destruction in the wild due to various factors. Since risk is caused by a series of events (escape, multiplying, damage), the risk similar to the risk chain shows that the strength of the risk is as solid as its weakest point. Therefore, if the probability of any link can be shown as close to zero, there is no need to quantify all risk aspects. Thus, the weakest link describes the upper risk limit. The first link is the possibility of harm when genetically modified seafood spreads, also known as the "so what" question, which is the toughest to determine.

Given the almost infinite possible exchanges within and between ecologies, it is clearly unrealistic to predict all possible harms from exposure. As per the current level of science, it is even impossible to predict the probability of potential damage, such as species migration or extinction. The second link, which is the possibility of escape, spreading and becoming wild, depends largely on the species. Transgenes that lower the sexual maturity age, such as growth hormone genes, will also be a key factor in population expansion. Seafood reaching a mature level will rapidly increase in quantity over multiple generations. Another significant factor is reproduction, which is usually stronger than practicality. If genetically modified male seafood is more appealing to wild-type females, the probability for Trojan horse genomes will increase (Zargar, 2020).

In nature, the mating ability is a more important evolutionary force than Darwin's survival for the fitting. If the species cannot be passed on through genes, what are the benefits of survival to sexual maturity? Relative fecundity (the amount of eggs spawned by female seafood, the number of sperm produced by males, motility and survival chances) determine the amount of offspring for each mating pair. Ultimately, the adult survival rate defines the rate with which the pair can replicate the mating process once, and each individual produces more offspring. These factors are interrelated, even when one of the factors is low, genetically modified seafood may still affect wild populations. For example, if juvenile viability is minimal but the mating success rate is high, then genetically modified seafood (if it can survive to maturity) can still spread the transgene over to wild-type species, thus reducing the survival prospects of these two populations.

If the possibility of transgene transmission in wild populations is significant, then efforts can be focused on biological techniques to manage the transgene, such as by using sterile seafood. This will also enable supervisors to understand whether all genetically modified seafood should be examined for sterility or just samples and the overall size of these samples. Since transgenes are typically implanted into domesticated species with poor adaptability, the transgene may be related to maladaptive genomes under natural conditions. While the knock-on effects between transgenes and other genes will disappear through recombination over time, they are still very high momentarily, which may challenge net fitness assessment. These problems can be solved by crossing the genetically modified seafood across the wild background, and casually breeding for some generations to remove the knock-on impact, and then estimate the ecological fitness (Zargar, 2020).

Nevertheless, this method does not exclude the need to evaluate other events (like natural disasters or human fault). However, such disasters rarely occur and cannot be proven within a normal time frame. Therefore, the emphasis must be placed on cautious prediction of the transmission of genetically modified genes as the initial step in risk evaluation. The risk assessment of genetically modified seafood has become

It is essential for regulators and decision-makers to determine possible environmental hazards and formulate preventive measures. By allowing supervisors to describe and use a regular set of standards, the model can provide a common method for evaluating genetically modified seafood as part of the standard risk evaluation process. Considering the general resistance to genetically modified organisms in general, this model (if widely approved by regulatory agencies and scientists) can make the risk evaluation process more structured and effective, and enable regulatory agencies to solve specific problems.

In addition, as shown in the political scuffles over the acceptance and regulations of genetically modified crops, risk assessment models widely accepted by regulatory agencies can prevent or resolve trade conflicts arising from the use and sale of foods produced from genetically modified seafood. Improved risk assessment processes will also help ensure public confidence in this possibly beneficial technology, which is urgently needed to address the concerns and increased resistance to genetically modified seafood in the United States and Europe. In view of the reduction of wild seafood stocks globally and the increasing pressure on natural resources, genetically modified seafood cannot be easily disregarded (Zargar, 2020).

Other possible pros and cons of genetically modified fish and seafood:

I. Advantages

Figure8.2: Labeling of seafood helps consumers to know whether their product is GMO or natural

Source: https://cban.ca/gmos/products/ge-animals/ge-fish/how-to-avoid-eating-gm-salmon/

Proponents of biotechnology insist that genetic modification technology has many benefits over traditional breeding techniques, such as, faster and more precise selection of positive traits. Since scientists can directly choose the traits they want to develop or augment, the required changes can be realized within a few generations, thus making it faster and cheaper than traditional methods, which may need many generations of choice selective breeding. Genetic modification technology allows researchers to precisely choose the traits to be improved, so that they can create organisms that are not only larger and grow faster, but can also be improved by, for instance, increasing nutrient content.

Proponents argue that faster-maturing fish can make fish cultivation more productive, increase production, and reduce the amount of feed required, thereby reducing waste. With the massive development of wild fish resources, genetically modified fish and seafood may be a significant means to meet the growing nutritional needs of humans and solve food security issues. Shellfish and finfish that have been genetically adapted to increase disease resistance can minimize the use of antibiotics. The increased cold resistance of fish may lead to the capacity to cultivate seafood in previously harsh environments and expand aquaculture to previously unconducive areas. Research is also underway to improve human health through genetically adapting fish to create human drugs like blood clotting factors, and producing shellfish that do not cause allergic reactions (DeSalle & Yudell, 2020).

Supporters of biotechnology claim that these benefits can be translated into various potential benefits, including lower costs for producers, lower consumer prices for seafood and medicines, and ecological benefits, such as reducing waste water pollution. Food researchers and the aquaculture sector may further support the inclusion of genetic technology. The premise of engineering is that product safety, environmental matters, ethics and information matters are satisfactorily resolved.

II. Disadvantages :- Contrarily, although most consumers seem to have largely accepted genetically modified food and feed plants, it is still unclear whether consumers will accept genetically modified seafood. Although the taste of this type of seafood may be the same, and like traditionally farmed fish, the price is expected to be cheaper than that of wild fish, but due to moral concerns on the proper use of animals and environmental concerns, this may affect public opinion and acceptance of genetically modified seafood.

The ongoing campaign by environmental organizations and consumer groups has required grocers, restaurants, and suppliers to sign a commitment not to sell genetically modified fish products even if approved by the FDA. Furthermore, the commercial fishing sector maintains that it has effectively educated the public to distinguish fish from different sources, like wild and cultivated salmon. However, the publicly announced escape of genetically modified fish may reduce public approval of their wild products. Environmental and consumer groups require special marking of genetically engineered foods. However, industry groups worry that such labeling may lead consumers to assume that their products are not safe to eat (Robert DeSalle & Michael Yudell, 2020).

8.3 DRIVERS OF DEMAND FOR GENETICALLY MODIFIED SEAFOOD

The US Food & Drug Administration (FDA) recently sanctioned the first edible genetically modified animal, AquAdvantage Salmon, which is a salmon type developed by AquaBounty Technologies about two decades ago. The FDA recognizes that this salmon breed is as safe to consume and as healthy as other wild or farm-bred Atlantic salmon, and they provide strict guidelines on where and how such fish are cultured. Although many ecological and seafood safety organizations and citizens object to this, GMO sea fish are still likely to show up in grocery chains and restaurants.

The aquaculture sector has grown to fill the space between the lack of wild seafood and the growing demand for consumer seafood. It has been proposed to use genetic modification (GM) technology to solve the sustainability issues related to existing aquaculture practices, but facts have proved that genetically modified seafood is controversial between industry investors and producers in terms of the requirements for genetic modification of food. Awareness of health and environmental risks, confidence and concerns about possible health and environmental risks, subjective awareness, and ambiguity prevention in the field of genetically modified foods are also important predictors of seafood selection. These findings have significant implications for food marketing using new food technologies. More so, people familiar with genetically modified technology are more inclined to eat genetically modified seafood or seafood fed by genetically modified organisms (Conceicao & Tandler, 2018)

Practical information interventions for users will include data on the health and environmental hazards associated with genetically modified seafood. As the worldwide population continues to expand, the demand for high-quality, safe and different foods, especially seafood, will continue to increase. Consumer views, attitudes and behaviors do play a significant role in determining consumers' preferences for seafood items. Risk perception, assurance, and anxieties about the risks related to genetically modified technologies are the main behavioral factors in the health and environmental fields. Context-based subjective understanding and ambiguity aversion both promote the consumption of seafood marked by the authorities.

Understanding GM technology is another important aspect of the demand for genetically modified seafood, and the ultimate efficacy of solving the problem of information asymmetry in the GM seafood market. In addition, there is evidence that the use of genetically modified technology in seafood will promote the intake of these products, so long as the consumer perception is positive. Nevertheless, concerns and confidence around the possible health and environmental dangers associated with genetically modified technology can collectively push consumers away from the seafood market. Although labels under seafood regulatory agencies can effectively promote consumers' informed choices, labels may inadvertently reduce overall seafood consumption. This is a problem considering the already low consumption of seafood products in European countries. This is especially important given the recent lifting of the import restriction on AquAdvantage salmon roe and the mass production of the product.

Based on these findings, future work will require the use of machine learning and other empirical technologies to better comprehend how behavioral measures can lead to adverse consequences of the newly proposed food policy. This is a more urgent matter. Some other considerations to examine as predictors of behavior may include the measurement of cultural perceptions and social norms related to science communication and public opinion. Consumer views on the sustainability of genetically modified seafood products are a particular area worth examining. If the progress of the aquaculture sector, including genetically modified technology, is applied to solve the sustainability problem, the ultimate effectiveness of such progress will depend on the perception of consumers. Therefore, additional investigation of consumers' opinions of the viability of industry innovation will be a valuable effort (Conceicao & Tandler, 2018).

These analyses shall continue to gather evidence to support the incorporation of "unconventional" data into models to increase prediction performance. Although machine learning technologies can be applied to examine consumer behavior, these techniques have some integral limitations, which have an impact on their use in policy assessment. From a practical point of view, the main limitations of these techniques are the estimated time and computational requirements. Another drawback of using seafood consumer measurement techniques is that it prevents the estimation of effective willingness to pay (WTP) methods. Because of potential selection bias, WTP estimates using this method are invalid. Selection bias is caused by the difference between choices made from a set of related variables.

Either way, these models are only as practical as the data accessible to decision-makers. The European Union (EU) organizes the annual Eurobarometer to evaluate public opinion on different topics, from confidence in national governments to consumer trends related to sea fishery and aquaculture produce. The triennial special survey focuses on the EU residents' knowledge of biotechnology, and different studies use publicly accessible data sets to estimate the public's perception of genetically modified seafood. In the U.S., the Pew Research Center performs similar public surveys and avails data sets for community use. There are various ways in which the information on seafood tags and external market details may have different impacts on consumers' purchasing trends.

Specifically, scholars reiterate the significance of measuring risk awareness in a multi-dimensional way. These results provide the most useful information for countries that are conducting public surveys (such as the European Union and the Pew Research Center), which assess the public's views on new seafood technologies (such as genetic modification) in the most appropriate way. A comprehensive understanding of decision-makers is mostly relevant for policy makers, as regulatory processes that encourage the sustainability of seafood structures are developing concurrently with sustainable food innovation. Policy developers can make good use of these data sources to resolve policy prediction issues related to seafood processing labels. Using this data with model selection methods can help clarify key behavioral considerations that are critical to consumer decision-making in the seafood industry. Given the existing policy interventions aimed at promoting the consumption of seafood products by influencing the external information accessible to consumers, these techniques are applicable in determining how these interventions are effective and, crucially, can help determine future policy actions or strategic seafood marketing tools (Conceicao & Tandler, 2018).

8.4 GMO SEAFOOD CONCERNS AND CONTROL METHODS

The possible impact of genetically modified organisms escaping from aquaculture facilities has attracted great attention from some researchers and environmental groups. A report by the National Research Council pointed out that genetically modified seafood constitutes the biggest science-based problem related to animal biotechnology. This is largely due to the inherent uncertainty of early detection of environmental problems and the difficulty of finding the problem and providing a remedy. Critics mention that genetically modified seafood may reproduce with their wild counterparts and spread defective genes. A study observed that seafood commonly used for experimental models can compete with unaltered species in a laboratory environment, and thus put forward the "Trojan gene theory." However, the offspring of this genetically modified seafood when mated with wild species are not healthy, lacking some physiological or behavioral characteristics, leading to the ultimate extinction of the improved population.

The ecological dangers of stocking genetically modified shellfish in nature have not been comprehensively examined because, due to their breeding and dispersion techniques, it is harder to control and isolate these organisms than to restrict most marine fish such as salmon. Even if fast-developing genetically modified seafood will not disperse their genes to wild animals, critics still worry that genetically modified seafood may destroy the ecology of the natural habitat by competing with local seafood for scarce resources. The escaped genetically modified seafood may harm wild seafood by increasing competition or predation. Furthermore, some people believe that genetically modified seafood, especially those that have been modified to increase their capacity to withstand a variety of salinity and temperature, may be harder or impossible to eliminate, just like invasive species. The effects of this competition will depend on various factors, such as the status of the wild population, the ratio of escaped seafood and particular genetic strains, including the native environmental conditions (Gjedrem, 2005).

For instance, critics believe that when non-GMO Atlantic salmon raised in inshore cages in the Northwestern United States and British Columbia escaped into the rivers, signs of the severity of this potential problem were noticed, and in some instances exceeded the number of wild salmon. Nevertheless, it is not clear whether genetically modified fish can survive in adequate numbers in the wild to cause permanent damage to the population. A study showed that when food supplies are insufficient, genetically modified fish have the capacity to harm wild species, however, the authors warn that controlled experiments may not truly reflect the wild conditions. Proponents of biotechnology believe that if genetically modified fish can escape, they are unlikely to survive in nature, particularly when they are raised in protected artificial environments and have not mastered how to evade predators.

Related issues

The main issues involving genetically modified seafood are whether they pose a threat to the ecosystem or human wellbeing, and whether they are ethical.

- Environmental risk

  The United Nations regards environmental factors as the biggest scientific concern for genetically modified seafood. Their reasoning is that there are many uncertainties about how genetic modification can affect seafood and the impact on the environment.

  The three primary factors mentioned in the environmental risk assessment are: I) The impact of transgenes on genetically modified animals-what are the expected phenotype changes and what are the probable unknown effects? II) Specific animals that have been genetically modified-some seafood are easy to become wild or invasive animals and have a past of causing environmental destruction; III) Receiving environment-the ability of the environment to withstand the impact of genetically modified seafood or recover from it will be different, plus the environment has different values to various societies (Zargar, 2020).

  Most of the environmental dangers of genetically modified seafood are similar to those of non-GM seafood. Any organism that enters the environment will affect local biodiversity through predation/grazing, competition, habitat changing, and hybridization with native species. Because of insufficient information for long-term and comprehensive field studies, the capacity to evaluate risks is still insufficient. The ability of genetically modified animals to affect the environment also depends on the particular environment. Ecosystems have varying levels of resistance to adverse effects. Compared with unstable marine communities, resilient communities are less affected by genetically modified seafood.

- Health risk

  Even though most seafood regulatory bodies consider environmental issues to be of paramount importance, human health issues related to genetically modified organisms in the food chain have caused widespread concern around the world. Part of the reason is news about GMO crops. Some have been genetically improved to contain herbicides, pesticides, and general antibiotics. People are worried that these toxins will affect them. Unknowing consumers believe that genetically modified seafood may also contain toxins. Another difference between flora and fauns is that in general, animals do not create natural toxins or anti-nutritional elements like many plants. Therefore, the scope of inadvertently and naturally occurring toxins in animals is relatively small.

While the dangers to human health are small, they exist and must be considered. Genetically modified seafood can reveal genes or gene products that have no history of safe usage in human diets, namely proteins. Risk management experts are considering a procedural evaluation of genetically modified seafood. The evaluation first determines the possible differences between genetically modified products and non-transgenic products, and then determines the nutritional and toxicological effects of the differences. When examining the food safety of genetically modified products, consideration should be given to the DNA constructs applied for the modification of seafood, particularly if the genome or promoter is derived from a viral source. Here, horizontal gene transfer and recombination (the combination of different DNA) may occur and cause the production of new viruses. The DNA fragments might not be fully digested by the human intestine and may survive within the gastrointestinal tract. These pieces can be absorbed by the intestinal somatic cells (Zargar, 2020).

In crops, some foreign genes cause allergic reactions. For instance, a gene in Brazilian nuts was added to soybeans, and consumers allergic to Brazilian nuts reacted negatively to the soybeans. In the fisheries industry, the most common genetic constructs involve growth hormone genes, not herbicides or pesticides applied to plants. Many genetically modified organisms tested for aquaculture can only generate more of their individual growth hormone.

The labeling of genetically modified products is presently a controversial issue. However, many nutritional, toxicological and allergen problems can be alleviated through this label. Therefore, from the perspective of human health, the risk is very small. One of the potential areas of concern is the future building of disease resistance. In theory, if genetically modified seafood is more illness-resistant, it may host new pathogens, some being human-borne or disease-causing (Zargar, 2020).

8.5 ETHICS AND ANIMAL PROTECTION

Is it moral to genetically modify seafood? To answer this question, it is important to define the meaning of "ethics" and create an "ethical framework" for evaluating the problem. The moral dilemma seeks to address this concern, but it is different from a value system that largely depends on the particular culture. Most ethical frameworks occur or can be formed by different jurisdictions, but this section discusses the framework partially used by the FAO Subcommittee on Ethics in Food & Agriculture. The elements of the framework stipulate that ethics include:

- Beneficial- towards people, organisms and the ecosystem. The use of genetically modified organisms that endanger humans, harm or threaten animals, or have negative effects on the environment can be judged as unethical.
- Justice and fairness-the use of genetically modified organisms or the creation of technologies that allow the use of such substances must be constrained by moral justice and fair treatment. In the case of native people, without their consent and no compensation for using their seafood resources, the problem of patenting genetic resources may exist. It is believed that the U.S. Food & Drug Administration may soon approve genetically modified salmon for sale in the country. Nevertheless, there is another rumor that the FDA may disapprove the cultivation of salmon in the United States. This may be considered an unethical decision since it does not handle the US and foreign ecosystems fairly.
- Autonomy-the right of people or groups to self-determination. This may also include the capacity to know how to make wise choices. In the area of crops, the control of genetically modified seeds by multinational corporations prevents farmers from accumulating seeds and forces them to rely on multinational corporations, thereby reducing farmers' autonomy. The failure to label genetically modified foods also limits the ability to access information that consumers require to practice their autonomy when choosing genetically modified or non-GM foods.

The ethical questions about aquatic genetically modified organisms usually center on whether humans can modify natural creations. Are people overly exercising autonomy? Nevertheless, humans have been changing plants, animals and their ecosystems for thousands of years. The growth of agriculture is considered as one of the significant aspects of civilization because it provides time and resources to enable humans to feed more humans and free them to develop art and science. Genetic adaptation enables humans to transform nature much quickly and to a bigger degree than before.

8.6 GMO SEAFOOD CONTAINMENT MEASURES

a) Sterilization :- There are various potential safeguards to solve these environmental problems that may be necessary. Sterilization is one of these control measures. Seafood producers may only need to obtain approval of sterile genetically modified fish for farming. Fertilized seafood eggs that have been exposed to temperature or pressure retain an extra number of chromosomes. The resulting triploid seafood have no normal reproduction features. Generally, triploid females are more sterile than males. Therefore, all-female species of triploid seafood are among the best ways to ensure non-reproductive populations of genetically modified seafood. However, there are differences between batches, and it is unclear whether this technique is applicable to all species.

In addition, opponents question whether the escaped triploid seafood can mate with wild animals. Triploid seafood have sufficient levels of sex hormones in some species to achieve normal courtship behavior, so they will potentially mate with wild ones thus reducing the reproductive success rate of wild populations. Other sterilization techniques are currently being studied, and research in this technique may increase the choices available. Critics of genetically modified seafood believe that no matter how small local populations are, the risks still outweigh the possible benefits of the technology, particularly where native seafood species are threatened or endangered.

b) Confinement :- In order to most effectively reduce ecological risks, the National Research Council (NRC) report on the biological restriction of genetically engineered organisms suggests that each individual has its unique bioconfinement plan. Similarly, since no particular method is 100% effective, the redundancy of bioconfinement can significantly increase the possibility of control, particularly if not combined with physical restrictions. Breeding genetically modified marine seafood in isolated onshore facilities rather than offshore or inshore pens may greatly reduce the risk of breaking into the wild.

Regulations and emerging issues

Figure8.3: GMO seafood can supplement the need for protein in human populations

Source: https://www.newfoodmagazine.com/news/116163/seafood-could-account-for-25-percent-of-animal-protein-needed-to-meet-demand/

A study by the National Research Council concluded that the food safety risk of genetically modified seafood is low to moderate. Because genetic engineering can introduce new proteins into food, there are fears that this technology may introduce previously unidentified allergens into the food chain, or it may introduce known allergens into "new" foods. Within the FDA, the Veterinary Medical Center is responsible for supervising genetically modified animals for human consumption, and its supervision of new animal drugs has the same authority. Furthermore, genetically modified fish must comply with the same international safety principles, which are applicable to conventionally farmed fish. Under the adulteration regulations, FAO has the right to remove a certain food from the marketplace or impose sanctions on it if the food poses a threat to public health. It includes food safety regulations for genetically modified fish. Scholars have discussed the extent to which genetically modified organisms are needed in aquaculture and food supply. With regards to exploring the potential of genetic modification, the response is usually positive. However, it is clear that there is a serious lack of knowledge in the wider population to make informed choices around the value of genetically modified organisms. The task of establishing and managing sustainable, responsible and scientifically reasonable measures has become more difficult. Meanwhile, due to frequent irresponsible and incorrect media treatment of genetically modified organisms, on the other hand, some authorities have published arbitrary statements concerning GMO seafood, for example, they claim there is no big difference in the labeling between genetically modified food and non-GM food (DeSalle & Yudell, 2020).

This extreme difference in attitude may arouse public opinion. Researchers have put forward requirements for risk appraisal and risk management procedures related to the use of genetically modified organisms in aquatic species. At least as a common practice in developed countries, if biotechnology produces healthy products in an environmentally sustainable manner, then the public may support biotechnology. The statement implies one of the most important factors-when considering the pros and cons of genetically modified organisms, is public opinion and the attitude of regulators. At least some individuals suspect that some modern sciences may also be impacted by ideas. Recently, some people have expressed resentment at the genetically modified breeds allegedly found in contemporary cultivars in traditional Australian seafood varieties.

Inferring directly from agronomy may be dangerous to aquaculture, but it is worth mentioning that in the United States, which accounts for the largest proportion of most genetically modified seafood, actions are being taken to establish a stricter regulatory framework. The Washington House of Representatives proposed several bills covering legal protection for producers, improving the safety of genetically modified seafood, requiring the labeling of seafood containing or produced by genetically modified organisms, solving GMO concerns in developing countries and assuming responsibility for damages also done. These measures seem to be at least partly due to the serious dissatisfaction of some seafood producers who use genetically modified foods.

However, it should be pointed out that the central position of some large multinational agrochemical businesses in providing seeds and agrochemicals has not been captured in the aquaculture sector, and the aquaculture sector is more fragmented. However, there are already signs that the industry is moving towards less and fewer companies, which may have an impact on the control and use of genetically modified organisms in aquaculture species. A few countries have formulated management policies for the cultivation, release, and diet utilization of genetically modified organisms, but the methods and restrictions used are very different (DeSalle & Yudell, 2020).

At the global level, the Cartagena Protocol on Biosafety of the Biological Diversity Convention provides an all-inclusive and strict framework for regulating the protection of biological diversity. The vast majority of countries are parties to the convention. What is unclear is the degree to which agreements applied at the national and local levels through domestic regulation and effective sanctions can be implemented. In its provisions, the "Cartagena Protocol" deals with the risk assessment related to the protection of biodiversity in detail. Some important and relevant basic principles include: a) Risk assessment must be based on specific conditions. b) The risks associated with living modified organisms must be measured with regard to the risks modeled by the unmodified organisms in the receiving environment" and the [comparability] risk assessment method which could be: I) Based on the assessment of the likelihood and consequences of the recognized adverse risks, the general risk postured by the living modified organism is assessed. II) Recommendations on whether the risks are tolerable or controllable, including determining strategies to manage these risks when necessary.

Therefore, there is a need to face a variety of stakeholders. Breeders and growers, multinational businesses, supermarkets, consumers, geneticists, politicians and media. Everyone has their own agenda regarding GMOs, but it is essential and reasonable to solve conflicts by obtaining and using complete, reasonable and accurate information fully and reasonably, especially when formulating appropriate regulations affecting genetically modified organisms (DeSalle & Yudell, 2020).

8.7 TECHNOLOGIES FOR INDUCING TRANSGENICS

Transgenic fish are mainly produced by microinjection into inseminated eggs or early embryos. The electroporation of sperm has further been proven to be successful in certain seafood, such as Chinook salmon. Liposomes have further been used as carriers. In order to use them in transgenic crustaceans and seaweed species, ballistic methods comprising of microparticles have been studied in Artemia. Using "Baekonisation", an electroporation technique with a flat electric field, to transfer DNA to fish embryos, this method seems to be very successful, but it has not been thoroughly used like other types of electroporation and microinjection techniques (Solé-Cava et al., 2013).

Recently, the use of embryo stem cells (ESC) has been advocated as a method of inducing transgenes. These cells are indistinguishable and remain totipotent, therefore they can be processed *in vitro* and then introduced into early embryos, where they can support the host germline. In this way, genes can be firmly introduced or deleted. Although ESC technology has achieved early success in mice, early precursor cells have been grown from Medaka and have shown many similar functions as mice, the absorption rate of seafood technology is still quite slow.

Studies have shown that 90% of host cell blastocysts transplanted using Mes 1 cells develop into maturity, and these cells integrate into organs resulting from all the three germ films and differentiate into different kinds of functional cells. Another case of a new and perhaps more effective method for gene transfer is using holotropic retroviral vectors. They can infect a variety of host cells and infect newly inseminated Medaka eggs with the reporter gene, which appears to be integrated into the whole germline of some P1 females. In chinnok salmon, after comparing retroviral infection with microinjection, the two techniques are equally effective in delivering the transgene to the egg, but the difference in the degree of gene expression was reported among the microinjection founders was greater.

Nevertheless, the use of retroviruses comes with certain problems. The microinjection technique is suitable for a relatively small number of manipulated organisms, while electroporation, sperm/liposome facilitated and bombardment techniques are more appropriate for large-scale treatment. The most common method of inserting transgenes in seafood aquaculture is microinjection. Nevertheless, the problem of transgene mosaic expression is common, and this leads to changes in the proportion of transgenic genotypes present in the offspring (Solé-Cava et al., 2013).

Integration sites

Although research in this direction is growing, the factors that determine the location of integration are still not well understood. Due to the unpredictable effects of uncontrolled integration on resident genes, it is especially important to obtain higher accuracy of integrated control sites. Scholars have tested special sequences called transposable boundary elements in other species that confer expression independent of the position of the transgene, or enhance the ability of the transgene construct to integrate with fish chromosomes. Initial results show that such elements in certain species cannot act as enhancers or increase the frequency of integration. However, judging from the expression of the CAT gene in F1 fish, it was found that some animals' boundary elements confer position-independent expression. Scientists have shown that the co-transference of retroviral integrase molecules and genetically modified DNA can quicken and increase the integration speed. In the future, more studies of this kind are needed to improve the success rate and controllable positioning of transgene integration (Beamish& Rothschild, 2009).

In terms of improving aquaculture production, food security and producing economic benefits, more than half of aquaculture genetically modified organisms can provide various services. There is no doubt that genetically modified organisms are useful in aquaculture, but their use must comply with the ideologies of the Cartagena Practice. Especially in terms of integration sites, the accuracy and efficiency of transgene induction technology may need some improvement.

The integrated sequence must not contain virus-based DNA, reporter genes and other genes not needed for the target phenotype. The risks posed by the use of genetically modified seafood must be analyzed and quantified in a practical and reliable way. Furthermore, there is a need to disseminate balanced and honest information about genetically modified seafood among decision-makers, aquaculture farmers and the public to avoid stigma (Beamish& Rothschild, 2009).

CHAPTER 9

CONSERVATION BIOLOGY OF FISHERIES SCIENCE

CONTENTS

9.1 INTRODUCTION

Conservation biology has developed into one of the quickest growing fields in biological sciences. Given that effective fish classification represents the health of the aquatic system, this is logical. The focus of conservation biology is to understand the ecological procedures necessary to protect biodiversity, with a focus on fish and marine species and their ecosystems.

Biological parameters that describe key factors of fish stock history, such as gender, age structure, growth pattern and sexual maturity (reproduction and selection) dynamics, are generally regarded to be basic information for evaluating fish population changes, fertility, and the productivity of natural resources under development pressure, or responding to environmental impacts to establish accurate conservation level of free-range stocks, and develop conservation and sustainable fisheries supervision strategies. However, sometimes it is difficult to obtain certain basic data needed to forecast the stock dynamics of vulnerable species only by visual confirmation (Bert, 2007).

9.2 THE FRAGILITY OF THE FISHERY ECOSYSTEM

Figure 9.1: Preventing overfishing is one of the ways to preserve healthy marine ecosystems

Source: https://www.sustainability-times.com/environmental-protection/a-shift-in-financing-can-make-our-fisheries-more-sustainable/

Ironically, despite the increasing interest in the protection of aquatic species in the past decade, the rate and degree of degradation of the marine environment on which these organisms rely have also increased. The main reason for the loss of marine biodiversity is still the rapid population growth. According to UN statistics, between 1990 and 1995, the global population increased by about 500 million to reach 5.7 billion and has continued to grow at a rate of roughly 86 million persons per year. The impact of population increase on the protection of marine biodiversity is not trivial, because about 60% of the globe's population lives within 100 kilometers from the ocean, and most inland metropolises are concentrated around lakes or river systems, which eventually flow into the sea.

Some forms of marine degradation caused by population increase include natural transformation and dilapidation of natural habitats, different kinds of water pollution, expansion of freshwater resources, introducing foreign organisms, and overfishing. For instance, it is believed that 950 million individuals use fish as their main source of protein; nevertheless, roughly two-thirds of the marine fish resources on the planet are overfished or have reached the limit of sustainability. Human demand for the globe's freshwater resources is also increasing. Although fresh water covers only 1% of the earth's surface, the growth in human demand for fresh water has rapidly surpassed the human population. There are different reasons why the aquatic environment is so prone to degradation. These factors include the complicated nature of marine ecosystems, the interaction between marine and terrestrial ecosystems, and the nearness of human populations to marine systems (Bert, 2007).

The productivity of different aquatic environments largely depends on the properties of water as a solvent, including its capacity to ionize dissolved compounds. Consequently, inland and nearshore oceanic environments are not just impacted by internal biogeochemical systems, but also strongly influenced by processes in the adjacent land environment.

Therefore, such aquatic environments are very productive, and humans have studied the productivity of this food, particularly fish. Nevertheless, the strong assimilation elements of water have also caused inland and oceanic environments to be used as a seemingly endless "sink" of communal waste generated by terrestrial biomes. In many regions, the capacity of marine systems to take in the trashes has been exceeded, and marine wasteland has been developed, which is increasingly unsuitable for most lifeforms. Some energy and matter flow from the aquatic environment to the terrestrial environment, but far less than the flow in the reverse direction.

Thus, for example, deforestation near streams in the Northwest Pacific Ocean can cause major changes in the surrounding streams' biome, such as siltation, increased algae growth in exposed river sections, and a decrease in the amount of wood chips in the river. In many cases, since coastal estuaries and lagoons are filled with sediment from the upper basin, the impact of logging is felt downstream. In densely populated regions, land changes have a particularly serious impact on the aquatic environment, since urbanization and agricultural expansion have led to key changes in marine communities. Since aquatic environments receive all kinds of human waste, this leads to quick and continuous degradation, often causing long-term effects on the fish stock populations (Lawrence & Hemingway, 2008).

Although many of the most changed environments are usually dominated by foreign species, many aquatic environments still retain most of their local species. Compared with most open ocean systems, man-made concentrated freshwater systems cause greater degradation of these systems. The continuous growth of population, urbanization, development, and the rising demand for freshwater assets for irrigation may lead to further reduction of freshwater biota. Reduced diversity in near-water systems often spreads to the sea, particularly to shallow coastal regions and coral reefs that contain most of the marine species. The capacity of the freshwater environment to absorb a large amount of abuse has protected the coastal marine environment from the secondary impacts of human civilization, but this protection capacity is quickly being lost. For instance, about half of the globe's coastal environments are endangered by human-based activities.

In Asia and Europe, development-based activities endanger 69% and 86% of coastlines respectively. Of all vertebrates, more than half are fish. About 24,600 species of fish have been identified, but when the world's fish species are well known, the total may reach around 28,500 species. This taxonomic diversity extends to 482 defined families. 80% (384) of the fish families have no more than 50 species, of which 14% (67) contain only one species. These statistics reflect the different adaptations of fish species to special conditions and lifestyles. Fish families consisting of just a few species include ancient lineages such as lungfish (three families, 9 species), coelacanths (Coelacanthidae, 1 species), and sturgeons (Acipenseridae, 25 species) (Lawrence & Hemingway, 2008).

9.3 FISH AS INDICATORS OF MARINE BIODIVERSITY

The primary reason behind using fish to check marine biodiversity is that humans know fish better than other marine organisms. As mentioned earlier, they are also fairly simple to collect and identify. Nevertheless, there are other convincing arguments for fish use. First, there are many kinds of them, and different species reflect different environments. Population growth and industrialization have caused most aquatic ecosystems to be polluted, leading to the deterioration of environmental water quality. The indicator organisms are required to improve the assessment plan on the environmental impact of human activities on the aquatic ecosystem. Due to the different sensitivity of fish species to pollution, fish have been extensively recorded as valuable indicators of ecological water quality.

The diversity, abundance and health of fish species have been assessed, and the relationship between fish communities and ecological variables assessed using procedures like Canonical Correspondence Analysis (CCA). The results show that urban sewage has a negative impact on water quality, and therefore on fish community structure. In a study of East African lakes, 29 fish types were recorded, of which the 15 most-affected species had the least number. Measurement scales like the Shannon diversity & Margalef species richness index show the highest readings at the reference point, and least at the most affected places. Wetland haploid species dominate the reference location, while oreochromis species account for the most affected. Compared with the offshore areas, the onshore areas have higher species diversity and lower species richness.

Minimal dissolved oxygen, secchi depth, pH and high electrical output, overall phosphorus and nitrogen levels are closely related to the sites affected by sewage, and greatly affect the structure of fish communities. The wetlands on the edge of the lake have basic social and ecological functions, which can maintain the livelihoods of rural and suburban communities, while contributing to coastal stability, flood management, water storage, cleansing and replenishment, nutrient preservation and microclimate regulation. Besides, lake wetlands and coastline shallow bays play a significant role in the protection of fish biodiversity. In particular, the rocky and low-lying inshore habitats are covered with aquatic plants, which are important refuges for vulnerable indigenous fish (Bert, 2007).

These densely-vegetated wetlands provide structural and low-oxygen refuges for fish that can withstand such conditions. They also serve to prevent the spread of certain fish, thus protecting cichlids from their predation. Nevertheless, the social and ecological functions of wetlands and the sustained survival of these species in these habitats are vulnerable to human activities in the catchment areas. Due to the increase in population, lake wetlands in developing countries are being widely degraded due to human activities such as, farming close to water sources, deforestation as informal slaughterhouses, carwash facilities, poultry houses and small-scale shops around the lakes.

These activities can be a source of point pollution and non-point pollution, and cause an increase in wastewater discharge, which carries potential pollutants and leads to the decline of the marine environment. This is caused by changing physical and chemical elements beyond the safety level, thereby affecting biodiversity, especially the biodiversity of fish. The fish resources and fish species diversity of East African lakes have dropped due to overexploitation, the use of poor fishing gear and techniques, environmental degradation, and poor management and destructiveness. The drop was also due to the inclusion of non-native Nile perch. The growth in the population of this species population coincides with the decrease in populations of many indigenous species. Long-term recommendations for dealing with this complex socio-ecological issue include establishing effective research and management plans, robust policy and legal structures, and information systems to manage fish and fishery habitats (Bert, 2007).

Direct interventions include law enforcement to control fishing efforts and using illegal fishing tools and methods, as well as supplementary programs and aquaculture that are still small-scale. Environmental changes caused by the discharge of industrial and household wastewater may reduce the entry of fish into wetland refuges in urban waters. It has become necessary to assess and mitigate the environmental effect on water quality and biodiversity. Endeavors to prevent or change the impact of the aquatic environment require effective and reliable detection mechanisms, especially system pollution detection mechanisms. Therefore, a biological monitoring system needs to be established. Biomonitoring studies endorse and recommend bio-monitoring as a valuable tool for monitoring water quality because it integrates the response to all pollutant combinations, thus describing the overall impact on the ecosystem and biodiversity.

Because of the sensitivity of fish to pollution, it has been widely recorded as a water quality indicator. Approximating the number of fish in a given area remains a basic theme of ecology. Species variety is linked to the functions of ecosystems and helps to understand the systems and impacts of environmental pollution. The evaluation of species diversity is equally useful for discovering trends, impacts or restoration of ecosystems, and quantifying extinction risks, so as to prioritize conservation of biodiversity hotspots. Preferably, a complete list can fully illustrate the abundance and diversity of species, but there are few. Instead, the sampling of biological population surveys provides significant data. Fish are perfect for ecological surveys.

They are easy to identify and rich in variety. In addition, warm tropical waters along with high visibility are also favorable to observation. The endemic haplocyanin cichlid species in the East African Basin is known to be the most varied and far-reaching radiation in vertebrates evolution. This makes bodies like Lake Victoria and their watershed areas important research zones for fish stock observation. Knowledge of water quality indicator organisms is essential for raising cichlids and regulating the loss of biodiversity in lakes. Nevertheless, although Lake Victoria is very important for water management and ecological monitoring, the connection between water quality and fishing stocks is rarely documented (Bert, 2007).

9.4 THE RELATIONSHIP BETWEEN FISH POPULATIONS AND ECOLOGICAL VARIABLES

The distribution and richness of fish species are affected by the water transparency, dissolved oxygen ratio, electrical conductivity and nutritional level (especially total phosphorus concentration) in the water. Among them, more species of fish have high dissolved oxygen concentration, high secchi depth readings and low conductivity levels. Similarly, the distribution and richness of fish are also affected by the quality of water. The closest research conducted in the Tanzanian area of Lake Victoria shows that fish species richness is significantly related to water transparency density. The low species abundance and even rocky habitats are credited to reduced water transparency.

Studies have shown that water transparency is among the strongest ecological predictors of fish species biodiversity, particularly in unimodal mineral species. The decrease in water transparency associated with eutrophication may have led to reduced species diversity of cichlids. It is reported that the transparency of water is highly correlated with the concentration of dissolved oxygen, because transparency allows light to penetrate, which is necessary for the photosynthesis process. It has also been reported that electrical conductivity is a key variable that can predict the form of fish aggregates, apart from water transparency and various other factors like alkalinity and macrophyte environments. Scientists also report that water depth affects fish diversity (Lawrence & Hemingway, 2008).

Figure 9.2: Dumping of raw sewage in the lake can threaten fish populations

Source: https://www.afrik21.africa/en/kenya-local-governments-take-the-lead-in-combating-pollution-of-lake-victoria/

However, the water depth has no significant effect on the distribution of fish species. Similarly, the water depth has no major impact on the species diversity and movement of small fish in Niushan Lake. Electrical conductivity and water transparency are effective environmental predictors of species richness in lake habitats. Areas with high levels of nutrition, particularly total phosphorus absorption, low biodiversity and richness, indicate that eutrophication reduces biodiversity. The total and relative abundance of carp are closely related to the nutritional status of the water body, particularly the total phosphorus content. Sites with low water quality have limited species richness. There is only one species of fish, *O. leucostictus*, which is highly related to eutrophication and can tolerate eutrophication. It is an important research fish for monitoring eutrophication in Murchison Bay.

Another closely related species that also dominates in other locations is Nile tilapia. Observations indicate that eutrophication and low water quality have a negative effect on species diversity. This is consistent with findings that eutrophication has a negative impact on fish biodiversity. There is a correlation between urban sewage discharge and eutrophication; this damages the ecological reliability of surface water, leads to oxygen depletion and adversely affects fish stocks and species diversity. Water quality strongly affects the distribution and abundance of fish. The evidence shows that human-made changes caused by ongoing urbanization of the catchment zone, such as the increase in sewage discharge into the environment, manufacturing affluence and municipal wastewater, have led to eutrophication.

Compared with the less affected locations, the more affected locations have lower diversity in fish community structure. Therefore, fish diversity might be a valuable biological measure of water quality that can be used in biological monitoring networks and programs to evaluate water quality and include fish hotspots. The scientists further suggest that to promote the sensible use and protection of fish, biodiversity and water purity; conservation efforts should include changes in social activities in the basin to improve environmental quality. For instance, promote the treatment of wastewater and sewage and biological radiation treatment from point source polluters like industrial plants and hotels before being discharged into the water body (Lawrence & Hemingway, 2008).

9.5 IMPACT OF FISHING ON SPECIES AND GENETIC VARIETY

Since the earliest single-celled organisms filled the earth, some genes have not changed. Only due to recent disease epidemics, genes exist in their current form. The conservation of genetic resources is a significant part of protection. Species loss in the marine ecosystem is not as prevalent as in freshwater or land systems. Nevertheless, knowledge about the number of aquatic species and extinction events is not perfect. Even in well-studied areas, other species are still getting discovered. The extinction of aquatic mammals and gastropods has been recorded (Bert, 2007).

Among them, overfishing led to the extinction of Steller's manatee and the depletion of Caribbean monk seals. Threats to upstream, estuarine and freshwater breeds and population extinctions have also been recorded. The amount of spawning constituents is a guide to help managers protect species diversity because they are usually recognizable in space and time.

Even in large numbers, fishing can change the genetic variety within a population. When fishing is exceedingly selective, it may permanently change the population characteristics, usually in the direction of lower economic value. Eliminating large fish seems to favor slow-growing, early-developing fish. At all biological levels, marine species may become extinct when conditions are unfavorable. The genetic structure of marine fish is much more than previously thought; selective fishing may lead to genetic differences in life-history characteristics and yield.

The cells of most living organisms have genetic codes that guide biochemical processes and growth. The code consists of 4 letters which are

C, A, U and G, which are transcribed in sets of three into 1 of 20 different kinds of amino acids, the popular protein "building blocks." The properties of protein affect different characteristics of the fish species, from temperature tolerance to skin color or body shape. The code translation is similar in all organisms from the tiniest fish to sharks. Although it has been established for billions of years, it was not revealed only until recently. On this basis, the diversity of organisms has greatly increased. The genetic string has another distinctive feature, which is replication. It is responsible for organisms that produce comparable models of themselves. This genetic code is the reason why a fish species produces other similar ones. For instance, Tuna can only reproduce other Tuna and not Cod or Sardines (Beamish& Rothschild, 2009).

Nevertheless, most advanced organisms cannot produce exact copies (clones) of themselves. Instead, the genetic codes provided by the mother and father are recombined to create a unique organism in their offspring. The way it happens follows the rules of genetics and provides a unique theoretical basis and genealogy for genetic data, which is unique from all other kinds of data collected. It represents a unique evolutionary lineage that cannot be directly replaced. Although some scientists believe that a fish species without any close relatives are more worthy of protection than a species with many close relatives, the truth is that the genetic combination in any fish species is unique. Maintaining the genetic diversity of marine species is a significant part of the conservation process.

To conclude, conservation biology has developed into one of the quickest growing areas in biological sciences. Since vibrant fish populations represent the good health of the marine ecosystem, this is logical. Some key aspects of conservation biology include; evolutionary biology, the legal basis of conservation, and public interest (Ray, 2018).

CHAPTER 10

THE FUTURE OF FISHERIES SCIENCE

CONTENTS

10.1 INTRODUCTION

Although technology is driving changes in almost every economic field, the global fishing industry of USD 130 billion is still not benefiting. But there's great potential to change this pattern. Equipping industry participants with powerful digital tools can help prevent industry losses while focusing on the greatest risk to our oceans. Of the nearly 10,000 fisheries in the world, only a small number use rich data sources and advanced science and technology. Most fisheries do not even monitor their catches. Those who do collect surveillance data usually use paper forms, and someone has to copy it into an electronic form and prepare it for analysis. This procedure can take months and is prone to errors.

The adoption of smart fishing practices through technology can produce unprecedented new information concerning fisheries and the seas in real-time. Currently, at least 10% of people in the world live on seafood. Correcting decades of poor management of fisheries may increase employment opportunities and bring around $80 billion in income to the international economy. It may also reverse the greatest risk to our oceans-overfishing, which has put one-third of the globe's fisheries at risk. Fisheries management based on modern science has proven to be a remedy. Adding advanced technology can make the supervision and fishing itself more resourceful. Fishermen can increase their profits while protecting the marine ecology (Beamish& Rothschild, 2009).

Figure 10.1: Fisheries science can determine stock numbers in a given area to prevent depletion

Source: https://ar18.iiasa.ac.at/fisheries/?doing_wp_cron=1609521746.31179 30889129638671875

Since the focus of fisheries research is on native, species-specific management matters, no formal evaluations of the long-term prospects of global marine fisheries have been conducted. Extrapolation of current trends means that the expansion of marine fisheries to deeper waters will have a serious effect on biodiversity, and global catches will decline. An examination of scenarios covering different social development options shows that it is now possible to reverse the negative trends that have plagued fisheries and at least partially rebuild the ecosystems it supports.

Generally speaking, stakeholders are not satisfied with the results of the fishery managing process. There are two main categories of interest in fisheries: fishermen are interested in catches, while others are interested in fish populations and the protection of marine biodiversity and habitat. There are fundamental conflicts between these groups in the distribution of catches and the impact of fishing on aquatic ecosystems. Due to overcapacity and other factors that induce "fishing competition", harvesters often put tremendous pressure on regulators.

However, when scientific knowledge is accessible to support the restriction and is well documented, the regulator typically wins. In order to achieve a sustainable future for fishery science, some of the topics that must be addressed include: a) Adequacy of scientific data for fisheries supervision- for population assessment, scientific information related to aquatic ecologies and protected species, and for socioeconomic data and analytical information. b) Using existing scientific data and advice to supervise marine fish and endangered species. c) Whether the scientific expertise available is sufficient. d) Strengthening high priority areas of fisheries scientific activities. e) Funding (Beamish, & Rothschild, 2009).

10.2 OVERVIEW OF GLOBAL FISHERY RESOURCES AND CATCHES

Currently, the world's total fish production is about 98 million tons, of which 82.5 million are marine living and 15.5 million are freshwater species, including aquaculture. Between 2008 and 2012, marine fish landings grew from 74 million tons to 86 million tons (an increase of 17%), while the corresponding figures for terrestrial water bodies were 10 million tons and 14 million tons (an increase of 38%). After reaching its peak in 2009, marine fish catches fell by 4% in 2012, while catches in inland waters continued growing, growing by 10% during the same year. Before, the increase in the catch of marine species was mainly due to the significant increase in the

North Pacific and South Pacific harvests, which grew by 17 million tons between 2001 and 2010.

The future landing volume will not continue to increase, and in overfished areas, production may decline. For example, in the North Atlantic and Mid-Atlantic regions, the fish catches during the period 2010-2015 remained almost the same (the annual change is less than 1.8 percent of the average), but the catches fell sharply in 2016 and 2017 (the changes were respectively 5.63 and 6.85%, compared with the average output of 2001-2010) (Beamish& Rothschild, 2009).

The type of catch is also significant because it depends on the market, that is, the kind of fish that customers want to buy. 13% of the total production is white fish, including cod, Alaskan Pollack, hake and haddock. Popular species, such as the only species in Europe, account for only 1.5% of total production.

The capture of sole fish is relatively stable. However, in certain Caribbean shrimp fisheries, the sole fish species on a commercial scale are often discarded as by-products. They are not highly regarded by local consumers, are cheap, and are often named irreverently. Due to the improvement of technical expertise and marketing level, such resources may gradually be integrated into the market. This development is essential to minimize discarded shrimp by-catch.

Most of the world's catches (about 30%) are composed of smaller pelagic fishes, which are small in size, full of lipids, and degrade quickly. Tuna, snoek, and mackerel, particularly the fatty species, are in greater commercial demand, accounting for 9% of the total catch.

Pelagic fish, particularly small fish are greatly influenced by environmental elements. This leads to changes in catches year by year. In Peru and Chile, most pelagic small-sized fish are processed for fishmeal production.

Overfishing is not the only reason for the decline in catches; natural factors such as "El Niño", environmental variations and longstanding biological cycles play an important role in some areas and have a significant impact on average landings. However, in the past decade, the amount of landfall has generally shown a stable trend. Therefore, the fishery sector has entered a period in which it is inevitable to fully manage fishery resources (Verspoor et al., 2008).

10.3 OVERVIEW OF CURRENT GLOBAL FISH UTILIZATION AND SALES

Seventy-two percent of the overall fish caught is utilized for direct human intake. Most of the rest (26%) is used to produce fishmeal (for raising livestock) and oil. Of all fish for human consumption, 31% are eaten fresh and 35% are frozen. Yet another 16% is processed as pickled fish (dried, pickled, smoked), and 18% is processed as canned fish. Therefore, 31% was consumed within the initial 2 weeks after capture, and 69% conserved in some way for later consumption. 39% of the catch is sold at the global level, and USD 37.4 billion worth of fish products is traded in some 199 countries. Developing regions play a big role in this trade. For example, among them are Mozambique and Cape Verde's fishery exports that comprise the primary source of non-aid revenue (Lawrence & Hemingway, 2008).

Fish discards and post-capture losses are high. Recent estimates indicate that the total annual catch and discards range from 17.9 tons to 39.5 million tons (an average of 27 million tons), which means that on average, it accounts for about 30% of the world's total catches. The biggest losses are experienced in selective fishing, like shrimp, where it is possible to find a bycatch/shrimp fraction of 5:1 or higher. Some countries, such as Cuba, strive to recover and use all shrimp by-catch, while others (such as Costa Rica) recover the majority of the valuable species on the market. Nevertheless, in many jurisdictions, all fish captured by shrimp nets are cast-off. Other losses are due to defects in fish handling, storage, supply and processing, and sales techniques. The actual amount of losses is a rather controversial issue, partly because it is difficult to define losses and measure them accordingly. However, all fishery parties concerned recognize that there are losses and must reduce the losses (Yáñez et al., 2016).

10.4 FISH TECHNOLOGY, ECONOMIC PRODUCTION AND FUTURE PROSPECTS

As a discipline, fish technology will continue playing a significant role in the development of fisheries. The design, processes and management of the fishery industry has undergone tremendous changes because of the synchronized introduction of quality check and management models, new mechanical and electronic equipment, the shift from physical to computer supervision and control, and the use of vision-based robots. The whole fishery industry will definitely change over the next few years: some of these modern fishery industries already exist in countries such as Iceland, Canada, and Denmark.

Due to this development, the connection between technology and economics is rapidly growing and will soon become significant in the future. For example, the daily observance of production economics was presented in some Scandinavian fisheries around 1975, electronic production control was introduced around 1982, while online product management in leading fisheries was present around 1995. Without the close connection between fishing technology and economics, this is impossible. This relationship significantly exceeds the classical relationship that exists between the two fields and is currently being used in investment research, because more specific and therefore more reliable factors (for example, output, time, temperature, process kinetics are required) (Reinecke et al., 2020).

As development seeks to grow profits and quality while reducing post-harvest setbacks and energy depletion, fisheries in developing countries must follow suit to avoid being compelled to withdraw from the market. There are fisheries industries, especially in Southeast Asia (Malaysia and Thailand), which have started this technology and have achieved substantial success. The expansion of aquaculture and moderate fisheries will bring further challenges to fish technicians, who are all linked to the economy. Fish as food, same as any other product, depends on supply and demand, however people, especially those in developing regions, are more production (supply) oriented rather than demand (marketing) focused. This may lead to poor investment and consequentl losses.

Salted cod and herring remained unchanged in the European market until the Americans began to produce frozen seafood in the second half of the 19th century, and ice was extensively introduced to cool the fish. Thereafter, because of the introduction of fresh technologies (management) or basically due to market development, various changes have occurred. Today, the European fish industry has few comparisons with the mid-19th century market. Despite the obvious and relatively rapid market changes, individuals in the fishing business in developing countries (and occasionally in developed countries) reason in a rather fixed and conservative way, which causes crises and economic damages (Reinecke et al., 2020).

A good model is the emergence and crisis of the cod industry in South America (intensified since then) by the lack of appropriate marketing data and opinions. Another very interesting illustration is the decline of the

Portuguese canned fish industry that was once the most influential canning business in the world, this happened due to poor technological development and proper market development. The latest market research shows that in the next few years, the demand for major food and fish markets will undergo many changes. These anticipated changes can be summed up as follows:

- Ordinary consumers, especially those in developed regions, will be older (the United States changed from 38 to 48yrs old around 2000). Consumers will therefore be more mindful and pay greater attention to safety, quality and nutrition. This trend of healthier fishery products shall also spread to developing countries, and may first start in some Asian or Latin American countries (such as China).
- Most people eat out regularly. For example, this is due to the rise in leisure activities as well as the rise in the number of professional women, including the fact that many individuals in big cities take lunch outside of the home during workdays. This means that catering markets shall increase, and products must adapt to these markets (for instance, size and weight, method of preparation). This trend is noticed in large townships in developing regions such as (Mexico City, Sao Paulo, Shanghai, and Lagos), where millions of individuals no longer return home for lunch.
- People in developed nations now spend less on food than before, but they need higher quality food and solutions than before. In general, in developing countries, food expenses are rising, and people expect higher quality and services than previously. This means that the rational usage of resources and prudent management have become critical in the food business, especially in the fishing sector.

Balancing or reducing the available fishery resources does not necessarily increase fish expenses, because certain consumers, especially those from the institution-based and catering industries, can easily switch to substitute products. Considering all trends, the future fish market seems to become more segmented and expanded. The fishery sector must adapt to this trend and respond accordingly to demand.

Large-scale fishery industries that produce commodities using a single or few species, or produce single or few products having low added value, will increasingly face unforeseen market changes, just like the cod industry in Latin America. In the near future, an increasing amount of fish landings shall shift from the commodity sector to value-added yields. Increasing urbanization in developed and developing nations will exacerbate the trend of value-added (occasionally called convenience) products.

All present market changes are challenging the current situation. Meanwhile, they create business prospects for people who understand the market and can use the appropriate technology and management. The production of fresh fish and to a certain extent frozen fish industries and products rely on supply, especially the relative input of raised and wild fish. The field of aquaculture is quite dynamic. It has not achieved its full potential, mainly due to the high cost of fish farming. Over the past ten years, improvements have been experienced, particularly in the farming of high-value commercial species like shrimp (Latin America and the Far East), and fish species such as sea bream, turbot, and salmon (Reinecke et al., 2020).

The crisis in the salmon sector, where the price of freshly farmed salmon has fallen to the range of sardines and anchovies in certain countries, clearly shows the marketing and commercial constraints that aquaculture yields still face in various parts of the globe. Salmon is not the only example, and today's farmed sea bream can hardly reach half of its price around 5 years ago before the growth of aquaculture. In addition, the increase in shrimp aquaculture production indicates that after a given level of production, rates may fall. At present, despite the impressive achievements of countries such as China, aquaculture is still not yet regarded as an economic protein source for populations. Aquaculture will continue developing and will use other knowledge areas outside of biology to produce products at relatively low prices. There are still issues that may prevent its development and must be resolved; including fish food, internal quality and production organization to achieve higher output levels. Certain countries with suitable coasts, weather and relatively affordable feed and energy shall have advantages, such as the situation in Chile concerning salmon farming. There are various factors that can impact the development of fisheries and aquaculture. The increase in gasoline costs, and the consequent increase in fuel costs, may make the cost of aquaculture more economical with regard to the fish caught, although the increase in fuel costs will also impact feed costs. The fuel cost may hinder long-distance fishing and some fuel-consuming fisheries in the future (Beamish& Rothschild, 2009).

Large-scale, continuous and sensible publicity campaigns can expand the market for reared fish species (such as salmon) to supersede current market restrictions. Although the aquaculture business has potential for development, it seems that there will always be a wild fish market to meet market demand for species diversity, regional and traditional food, and gourmet food. Cured and canned items will continue playing a significant market role, particularly in developing countries where there is no refrigeration equipment necessary for the supply of fresh and refrigerated fish. There is no doubt that the marked improvement in living standards will change. Global trade will remain active because the needs of all parts of the world, particularly Asia, and even Western and Eastern Europe will still be unfulfilled (Beamish& Rothschild, 2009). The lack of common fish types will encourage the emergence of fresh products with higher added value, and the number of fish that are purposefully reduced to the fish meal may decrease. Given large amounts of discards and by-catch losses, efforts shall be made to reduce discards and by-catch (for example, through selective fishing gear) and take advantage of those unavoidable losses. Projection is not yet a safe field, and any prediction founded on current conditions or similar inferences (for example, the development of aquaculture in accordance with agricultural and animal husbandry models) is not necessarily correct. It is impossible to obtain historical predictions that identify the rhythm, trend, pattern or historical policy, and the future is essentially open. If fishery stakeholders are willing to learn from their own mistakes, especially to improve a particular situation, they can learn from the past, recognize which kind of activity is rational and conscious, and which activity is the undesired consequence of poor fishing activities.

- Reliable scientific data for fisheries management

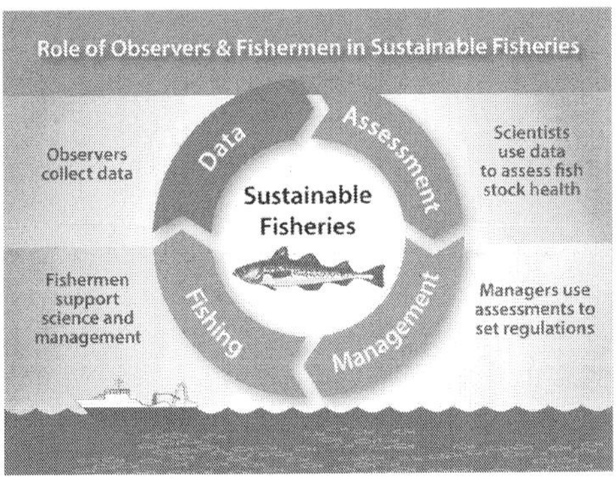

Figure 10.2: Sustainable fishing practices can be achieved through scientific research

Source: https://www.researchgate.net/figure/The-role-of-observers-and-fishermen-in-sustainable-fisheries_fig1_332976951

The data on the characteristics of fish populations gathered by fishery institutions are usually sufficient to guide the management of fish species, especially those of key socioeconomic significance. If it is necessary to analyze all fish stocks during fishing, the funds used to collect and analyze fishery data are relatively small compared to the arduous nature of the work. Given the present knowledge level, conservative single-species conservation is the most significant (and most cost-effective) method for most fisheries.

By increasing expenditures on data gathering and analysis (including observer programs), certain fisheries can be assessed, although more commercial data and information gained from cooperative surveys are used. However, for some fisheries, due to the widespread phenomenon of diminishing returns, the evaluation accuracy of data collection and analysis and the increase in the accuracy of each additional fee may be a diminishing function on investment.

The fishery management agencies must use different models (based on the available data) to examine the same data to better understand the data and discover the peculiarities caused by the assumptions implied in the model instead of the data itself. In order to achieve this goal, it is necessary to train stock valuation scientists to use different models more widely, and it is unlikely that the model will be used as a "black box." Other significant suggestions made by these fisheries scientists must find ways to utilize fishery-related data more efficiently and gather and use additional socioeconomic data in the population assessment process to assess the socioeconomic effects of different fisheries management strategies (Beaumont &Hoare, 2008).

All countries cannot evaluate fisheries to the same standards. Stakeholders should continue focusing their resources on measuring the most economically and environmentally important species. An all-inclusive cost-benefit inquiry of fishery data gathering and stock assessments nationwide will help determine funding priorities. Most fish stock evaluations are only performed every few years, and each evaluation is reviewed. For most Atlantic fisheries, peer assessment is a slow and rate-limiting procedure, forcing the management to apply outdated assessments. For other stocks, peer review is only required when a significant change in stock position is noticed or a key change in management methods is suggested. Fishery management relies on the timely provision of various biological, ecological, and socioeconomic data, plus stakeholders must participate in various activities to gather and manage such information.

Stakeholders mention that over time, reliable indicators show that fish populations are very important for fisheries management. For most fisheries, dependable indicators come from surveys that are independent of the fishery. However, surveys are usually carried out on relatively old fishery research vessels with outdated technology. It is recommended to use a more advanced survey fleet (especially trawl and acoustic survey ships), and replace the aging fishing equipment with newer, more powerful and quieter vessels. Governments should not only provide funds for the production of new vessels but also provide sufficient funds for the investigations and research carried out by these vessels (Beaumont &Hoare, 2008).

Data from surveys and other sources unrelated to fisheries are stored in various localities and formats with comparatively low coordination, which makes it hard for managers and researchers to access the information. Heading forward, there is a need to consider the development of a Global Fisheries Information System (FIS) plan. FIS framework (based on a synchronized regional system) is perfect, and its dependence on global standards is a progressive feature. FIS is ambitious; nevertheless, to succeed, (1) the government must provide sufficient funds; (2) it must ensure cooperation and coordination between regions. Governments must provide funding for the continuous acquisition and utilization of new vessels and fishery information systems (FIS) to better coordinate the fishery sector.

- Marine ecologies and protected species science

The fishery authorities are responsible for managing various legislative tasks and require a wider range of scientific knowledge than organizing traditional scientific activities. In its scientific work and the emphasis on legislative authorization, fishing agencies seem to pay more attention to regulatory agencies. They seem to carry out scientific activities on the basis of traditional fisheries-oriented priorities, but lack the scientific knowledge to meet the important legal requirements arising from their activities. Most fisheries are in a state of reconstruction and are trying to keep as close to (or exceeding) rebuild the stock to a level that can produce the best yield in 10 years.

Over the past decade, the environmental conservation community has been under increasing pressure to expand data on protected fish species and basic habitats. However, regulators still seem to focus the majority of their activities on protecting the fish harvest. They are liable to enforcing various laws, which may sometimes contradict goals and regulations, without clear guidelines on how to keep balance. Thus, protecting biodiversity, maintaining marine food chains and protecting habitats are significant goals that must further be incorporated in fisheries management. Development towards the fisheries ecosystem will add new requirements on fishery managers. As people gain more and more understanding of how different fish populations interrelate as part of the marine ecology, people increasingly realize that fish landings are not a proper measure for the health of marine resource (Beamish & Rothschild, 2009).

Regulatory agencies have the ability and facilities to derive observations and perform necessary experiments through their scientific centers, and relationships with educational scientists to improve the biological comprehension of fish stocks and marine ecosystems, thereby fully improving fisheries and protected resources such as sea mammals, sea turtles and seabirds. Nevertheless, in recent years, many scientific functions have been dedicated to gathering and evaluating data for population assessment, and other duties directly related to temporary needs to fulfill fishery regulatory requirements. This reduces the investment in finance, facilities, and human resources, making it impossible to conduct basic research, which is necessary to achieve present and lasting resource management policies. Besides, under constantly changing ecological conditions, such as climate change, environmental and biological research will become increasingly important.

At present, in terms of requirements and availability, the relationship between predator and prey, and the health and reproductive condition of the sampled fish stocks, there is limited data and research necessary to perform the tasks of regulatory agencies. A balance must be struck between the traditional emphasis on viable development and the role of protecting vulnerable species and environments created by fisheries. A fresh focus will require improved research on the biology and bycatch ecology, fish habitats, aquatic ecosystems, and threatened and endangered fish species. Fishery managers should look into gaps in scientific knowledge and legal loopholes when determining future research priorities (Beamish &Rothschild, 2009).

New forms of information must be collected and acquired from diverse agencies to ensure that fishery managers can explain the location of target species in the aquatic food web, the impact of fishing on aquatic ecosystems, and the impact of changing environmental factors on fish populations. It is necessary that those working with protected resources, ecosystems, and sustainable fish resources must communicate efficiently to plan, coordinate, and perform the required research. Ecosystem research may cost a lot of money and must be considered in financing priorities without reducing support for regular stock assessment and biological studies. Fisheries managers must consult to find out which lack of funding and behavior may cause the agency's greatest legal vulnerability. A policy to address these vulnerabilities is required.

- Reliability of socioeconomic data and analyses

The socio-economic data gathered by stakeholders is insufficient to understand the impact of former management on fisheries and the fishing communities, and to predict the results of alternative management options. Fishery management options often do not contain appropriate socioeconomic goals. Apart from collecting and analyzing biological data, fisheries management also requires socioeconomic anthropological inquiry of how participants react to and comply with management regulations, including how the regulations impact their livelihoods and overall welfare. The lack of fisheries socioeconomic data impedes the development and execution of acceptable and operative management measures. Insufficient data is caused by different factors (for example, insufficient funds, stakeholder restrictions, and fishermen's concerns regarding the confidentiality of the socioeconomic data they provide), which are basically beyond the scope of fishery managers.

The socio-economic objectives of fisheries are not openly and quantitatively expressed. Therefore, important socioeconomic values will not be continuously recorded or measured, so it is difficult to measure the progress of achieving goals, predict the impact of alternative management processes on socioeconomic values, and estimate the average cost of alternative management measures. Without key baseline data (such as data obtained through standard and regular data collection to monitor socioeconomic changes), it is hard to examine the impact of fishery management plans on commercial fishermen, recreational anglers and their related communities. Since important socioeconomic data is not collected and examined on a regular basis, it is hard to find out whether population fluctuations calculated by landings, for instance, are caused by variations in stock numbers or changes in similar socioeconomic factors (Shimizu et al., 2013).

Without sufficient socioeconomic data and analysis, it can be difficult to determine the overall cost of a management program. Thus, it is hard to reasonably discuss ways of minimizing the total cost and moderate the distribution impact as the public's need for the utilization of fishery resources keeps changing. There must be adequate information to develop plans that can reduce fishing capacity, while restoring the economic viability of fish harvesting, and keep the total harvest within the range needed to rebuild the population within a feasible period of time. Usually, fisheries managers make choices among alternatives based on biological considerations without accurately considering the socioeconomic impacts. These agencies require the calculation of populations relative to measurable definitions of unexplored areas and overfishing but do not cover similar requirements for socioeconomic values.

Obviously, overfishing capacity is among the main problems of fisheries worldwide, and there is not enough energy dedicated to collecting data and performing analysis to determine the overcapacity of every fishery. This information is necessary for formulating plans to reduce overcapacity, which is an essential step in minimizing risk-based management decision tendencies and decreasing bycatch and overfishing. The significance of socio-economic data and analysis for marine fisheries management must be determined in the fisheries business, the resulting federal guidelines, fishery management policies, funding applications and government grants. Different reports recommend increasing the collection of socioeconomic data. When formulating a fisheries management plan (FMP), it is necessary to consider socioeconomic and biological data (Shimizu et al., 2013).

In view of the fact that fishermen are using public trust resources, the policy should be revised to require the collection of such data. The availability of this data will allow fisheries managers to pick from different possible management options in order to balance biological, socioeconomic factors. As the authorities absorb new fishery scientists, their first role must be to quantify the excess capacity in all fisheries supervised under Fishery Management Plans (FMPs), and developing a plan for excess capacity that considers the socioeconomic attributes of individual fisheries. Plans to reduce overcapacity must consider all fish stocks available in a fishery area, because as fishermen move target species, the reduction in excess capacity in some fisheries may spread to other fisheries (mainly in the same area).

Governments must promote this kind of analysis because solving the problem of overcapacity can make fisheries management simpler and cheaper. Other tasks of high priority are to reach a regional and international settlement on the standards used to address social impact evaluation and global standard requirements, and to create a regional or global system to standardize and collect relevant social, population and other data on the socio-economic, cultural and community factors of marine fisheries.

- Using available scientific data and recommendations to manage marine fisheries and endangered resources

Overfishing may be due to a lack of adequate scientific data or the fishery management process ignores existing scientific advice, or lack of sufficient resources to impose management regulations. For the former, preventing overfishing depends on improved data collection and inquiry. In the last case, preventing overfishing depends on changing procedures to use science in the managing process and allocate resources for law enforcement. The governance system established by the stakeholders, and the disparity between institutional experts and responsibilities hinder the use of science in the decision-making process of marine fisheries management.

Regional fishery management agencies occasionally disregard the scientific advice offered by scholars in determining total allowable catches (TACs), and deciding other elements of fishery management practices (FMPs). The whole process is usually subject to extreme political pressure, straight from stakeholders and ultimately through fisheries lobbyists. Various issues emerge in how research is used to supervise marine fisheries. Stock assessment researchers can make clear recommendations about target fishing mortality and harvest levels, and the local councils must contemplate the recommendations in creating their FMPs. Commencing an appraisal of the fisheries management system and applying research in governance is essential.

Members of regional fisheries councils need to be aware of manageable uncertainty, and quantify the uncertainty, incorporate it into fish population assessment forecasts, and use it to make informed decisions. The board should explore various techniques to provide relative uncertainty and influence decision-making. Council members do not need to know the details of the stock valuation methods that apply the results of the valuation, but should help them evaluate the relative merits of criticisms of stock valuations that may emerge in the decision-making procedure, for instance, to represent the region at the Global Fisheries Management Conferences through a strong assessment of the fisheries department. It is essential to keep fisheries science from the political influence of supervisors. There may be a mismatch between the fishery researcher (advising the regional director) and the management staff (Shimizu et al, 2013).

Fisheries stakeholders must design and apply objective decision-making processes. Better organizational structures for scientific activities in fisheries will improve outsiders' insights of scientists. A structure that allows scientists to be objective and independent of the operation of management agencies can enhance the image and productivity of fishery management. Among the challenges and discontent with the fisheries management process are tied to scientific results considered suspicious, insufficient or erroneous. Different structures in the organization (which can better separate science from the managerial decision-making process) can improve the capacity of stakeholders to conduct science and improve the research quality that can be used in fisheries management.

Fisheries science must be clear and transparent to everyone, but also scientists must provide the best research work in an environment that is free from the political pressure that may arise in the management decision-making process. One of the ways to describe the management process is to make basic scientific input in the initial phase of the appraisal process, where the researchers present the evaluation results to the fish stock evaluation review committee, and the scientific and statistical agencies. But by then, most of the tasks on scientific advice should have been completed, with few opportunities for new analysis in the decision-making process. Therefore, new analysis methods must be developed at or before the original stage of the assessment process. This problem may be partially solved by enhancing the conditions for performing scientific fishery research. Fisheries researchers can benefit from workplaces where they can practice the following:

- Responsive science. Researchers must respond to the problem of resource management. For this, they need financial and human resources to respond. Preferably, scientists should realize that they must produce results that meet management needs in a timely manner, so as to get rid of the political pressure of managing related issues.
- Innovative science. Innovative science is needed to improve fish population assessment and reduce the staff requirements, costs, and risks associated with stock assessment and modeling. There is also a need to solve new and complicated issues in ecosystem science.
- Integrated science. New requirements for ecological science require a wide range of interdisciplinary methods to address climatic, oceanographic, the biology of individual species, system ecology, and fish population dynamics. An effective team is needed to perform such research to support management. In addition, fisheries agencies must cooperate to solve a wide range of ecosystem issues facing marine fisheries management.
- Visionary science. The science of foreseeing management factors is needed. To respond more to current problems, knowledgeable

staff and funds shall be required to carry out research with contemporary theories, modeling and expectation of fishing or ecological variability that will assist fisheries management. Some people believe that other institutions (such as academic institutions) must bear the role of research and science; nevertheless, if this factor is established, the condition of the fisheries science, the morale of the scientists, and the quality of the recommendations to the managers can be affected.

An environment that supports innovation and rewards brilliance needs to be created. Cooperation provides an environment for the development of innovative science through the joint efforts of agencies and academic researchers. Nevertheless, only a few scientists participate in these fisheries groups, and governments must recognize the significance of putting more fishery scientists into the environment.

Scientists recommend a series of surveys to help determine the reasons for the decline of this species and its interaction with fisheries, however, the research is usually not ranked highly in budget requirements and thus will not be applied within the period that makes the fisheries data available to the public. Similar issues have occurred with regard to the fish population identity of marine mammals. These issues are often planned for production reduction under the management of the fisheries supervision team. The research system must be able to perform research outside the ordinary multi-year budgeting plan and funding cycle because any delay in providing research results to management will considerably reduce the chances of avoiding species extinction. Besides, consideration must be given to growing restoration and protection plans to include the restoration of other vulnerable or endangered aquatic species.

10.5 EXPERT SCIENTIFIC INFORMATION AVAILABLE TO FISHING STAKEHOLDERS

Figure 10.3: Scientists measuring the health status of a fish habitat

Source: https://fishbio.com/field-notes/communication/supporting-women-fisheries-science

Existing scientific expertise mainly focuses on fish population assessment science and fisheries biology, and the demand for ecosystem science, protected species biology and social science is also increasing. Most fishery science activities are carried out internally. The academic group of marine fishery researchers is relatively small, and the supervisors have strong capabilities in fishery biology and fish population dynamics. The current staff's expertise is sufficient to enable agencies to continue emphasizing single species stock assessment. However, it may not be suitable for the data or research needs of additional ecosystem-oriented management, or meet the fairly new requirements proposed by fisheries supervisors (Hart & Reynolds, 2008).

However, it is not suitable for the data and scientific needs of more ecosystem-based management, and it cannot meet the fairly new requirements proposed in Global Fisheries Amendments, and the new comprehension of the fisheries requirements. Supervisors are often not fully prepared to deal with new expectations and new requirements about habitats, non-target species, endangered species or non-exploited use of marine resources because the budget is insufficient to fully expand their expertise in these fields. A plan for hiring and deploying social scientists among regions, scientific centers, and regional fisheries management agencies has been developed, but has not yet been completely funded and implemented (Hart & Reynolds, 2008).

10.6 FISHERIES SCIENCE RECOMMENDATIONS FOR FUTURE SUSTAINABILITY

Since scientists usually do not have time to perform basic research about fisheries missions, it is essential for agencies to maintain close contact with academic researchers through joint missions and external funding temporarily. Academia has developed many innovative concepts and technologies to support scientific fisheries missions. Researchers have also established joint and cooperative institutions, and located their regional fisheries science centers near key marine science institutions. This strategy has led to good cooperation between researchers and academic fisheries scientists (Beaumont & Hoare, 2008). Scientists have also begun implementing scholarship programs for fisheries graduate students to address the imminent shortage of fish population assessment researchers and resource economists. Fishery managers must intensify their efforts to draw new employees, particularly those with quantitative, financial and social science abilities, while retaining existing employees. These goals will need to rely on the historical distinction of fisheries science and increase the morale of institutions by increasing monetary and non-monetary enticements.

10.7 STRENGTHENING THE FOCUS OF FISHERY SCIENCE ACTIVITIES

Science has the strongest impact in basic fisheries biology and stock dynamics. Currently, stakeholders are doing important but relatively minor research work related to comprehensive bio-economic analysis, climate impact on fish populations, fish lifestyles in ecosystems, and the interaction between habitats and fisheries. A large part of the science budget is set aside for specific tasks, sometimes not related to fisheries. Although some tasks may have been carried out without designated purposes, they may have been provided by lower levels of funding and the overall quality is reduced. Basic funds can be used to adapt to new research priorities. The scientific priorities determined by the governments through earmarked funding are unlikely to be evenly distributed between the actual scientific needs based on the legal requirements and research programs (Beaumont &Hoare, 2008). The field of fishery science deserves more attention. The following is a list of various scientific topics that are considered insufficient, which may have contributed to the increase in litigation over the past few decades. It may be essential to redirect or increase the budget to support these activities.

i) Develop research plans and analytical practices related to fisheries factors to provide the information needed to protect endangered and vulnerable species and reduce the extent of lawsuits against fisheries.

ii) The collection and inquiry of spatial data and developing spatial population assessment models, so that the necessary information will be used in spatial management techniques, like designating marine protected zones, and incorporating knowledge into spatial fish population assessment models. With the emergence of definitions of fisheries and the requirements of most councils wishing to use different types of spatial management, new emphasis must be put on collecting and examining fishery data in the context of space. To obtain positive spatial data, a lot of work remains to be performed. One promising method that is broadly used is the Vessels Monitoring System (VMS), which helps link the capture location to the amount of capture in the vessel logs. Traditionally, although spatial population assessment models were available, fishery scientists have largely built models that include temporal characteristics but not spatial characteristics.

iii) As more attention is paid to multi-species interactions and ecosystem impacts, new models that include multi-species interactions and nutritional structure and environmental variability on fish populations should continue to be developed.

iv) Creating methods to link socioeconomic data with biological information in modeling and other different analyses. By revealing how abiotic factors affect catches per unit of work and other variables that are important in stock evaluations, such models should help make fishery-dependent data more beneficial in stock assessments. These models are useful for predicting the socioeconomic impacts of potential management options, including how they affect different stakeholders.

v) Gradually understand the size, circulation and time pattern of the fish industry and non-industry value changes with diverse management solutions.

10.8 FISHERIES SCIENCE FUNDING

Fisheries researchers are faced with the daunting role of facing pressure to maintain the maximum sustainable yield of most fisheries, fulfilling the new requirements included by the amendment including the pressure to fulfill these requirements, which require lots of funds to hire a strong labor force or maintain traditional data collection activities for fish population assessment activities. The basic budget is fairly stagnant because budget growth can be offset by special grants which may not meet scientific priorities. The impact of designated uses on scientific priorities should be studied. For instance, it is essential to define the percentage of scientific funding for specific scientific factors compared to the ratio of earmarked funding. This can be done by equating the fisheries budget request with actual appropriations.

Governments must check the data collection, analysis and management costs required by fishery researchers to complete their current tasks. Stakeholders in regional fisheries management often develop plans that require precise estimates of the size of fish stocks and the impact of alternative managerial options to avoid management failures. Most of these plans lack a proper buffer against uncertainties in fisheries science. The cost of such data gathering and management has not yet been fully analyzed. For instance, the cost of being fully covered by observers in a fishery where the bycatch rate is unknown must be determined. Similarly, the cost of the entire vessel monitoring system (VMS) should be factored. In conclusion, there are various trends in the development of fisheries, which in practice may help fisheries to truly achieve a sustainable future, such as reducing pollution and preventing overfishing. In the context of rapid population growth and increasing environmental pollution, among all the problems faced by the fishing community, the future of fisheries is affected by political trends and dominant socioeconomic factors. The expansion of fisheries is facing the need to provide for an ever-growing population, and is determined by economic interests and limited biomass, which has caused consequences for the depletion of developed populations. Economic elements, social elements, and fishery dynamics face one another and have multiple issues affecting each other, however, each has its unique dynamics, making the likelihood of finding isolated and autonomous solutions more complicated.

Aquaculture seems to bring hope to the fisheries sector by helping to solve the concern of generating food from the marine environment. Nevertheless, this is not a lasting solution. There are some trends in the development of fisheries, which in practice may help fisheries to achieve a truly sustainable future, such as reducing emissions, gradually reducing the output of large vessels, improving management measures, and increasing participation in the fisheries sector. Eco-labeling of fishery products, minimizing illegal fishing, ensuring price stability, and certification of fisheries are beneficial. All these factors have opened a window of hope so that people can expect the sustainability of fisheries to become a reality, not a farfetched idea.

REFERENCES

1. Andrew J. Lawrence & Krystal L. Hemingway. (2008). *Effects of Pollution on Fish: Molecular Effects and Population Responses.* New Jersey: John Wiley & Sons, 1-83.
2. Andy Beaumont, Pierre Boudry, & Kathryn Hoare. (2010). *Biotechnology and Genetics in Fisheries and Aquaculture.* New Jersey: John Wiley & Sons, 1-99.
3. Antonio M. Solé-Cava, Claudia A.M. Russo, & John P. Thorpe. (2013). *Marine Genetics.* Berlin: Springer Science & Business Media, 15-91.
4. Bert, T. M. (2007). *Ecological and Genetic Implications of Aquaculture Activities.* Berlin: Springer Science & Business Media, 1-85.
5. Biljana Vucelić-Radović, D. L. (2019). *Application of Molecular Methods and Raman Microscopy/Spectroscopy in Agricultural Sciences and Food Technology.* London: Ubiquity Press, pp218.
6. Bozkurt, Y. (2018). *Recent Advances in Zebrafish Researches.* New York: BoD – Books on Demand, 27-83.
7. C.-S. Lee & E.M. Donaldson. (2012). *Reproductive Biotechnology in Finfish Aquaculture.* London: Elsevier, 3-99.
8. Dunham, R. A. (2011). *Aquaculture and Fisheries Biotechnology and Genetics.* London: CABI, 1-93.
9. Eric Verspoor, Lee Stradmeyer, & Jennifer L. Nielsen. (2008). *The Atlantic Salmon: Genetics, Conservation and Management.* New Jersey: John Wiley & Sons, 1-57.
10. Farrell, A. P. (2011). *Encyclopedia of Fish Physiology: From Genome to Environment.* New York: Academic Press, pp2272.
11. Freeland, J. R. (2006). *Molecular Ecology.* New Jersey: John Wiley & Sons, 1-63.

12. Garth L. Fletcher & Matthew L. Rise. (2012). *Aquaculture Biotechnology.* New Jersey: John Wiley & Sons, 21-81.
13. Gary R. Carvalho & Tony J. Pitcher. (2012). *Molecular Genetics in Fisheries.* Berlin: Springer Science & Business Media, 2-55.
14. Gavin Burnell & Geoff Allan. (2009). *New Technologies in Aquaculture: Improving Production Efficiency, Quality and Environmental Management.* Netherlands: Elsevier, pp1224.
15. Gjedrem, T. (2005). *Selection and Breeding Programs in Aquaculture.* Berlin: Springer Science & Business Media, 1-88.
16. Graham Rowe, Michael Sweet, & Trevor John Clark Beebee. (2017). *An Introduction to Molecular Ecology.* London: Oxford University Press, 1-91.
17. Gupta, P. K. (2008). *Molecular Biology and Genetic Engineering.* India : Rastogi Publications, 1-62.
18. José Manuel Yáñez, Ross Houston, & Scott Newman. (2016). *Genomics in Aquaculture to Better Understand Species Biology and Accelerate Genetic Progress.* Switzerland: Frontiers Media SA, 6-81.
19. Kim, S.-K. (2019). *Essentials of Marine Biotechnology.* Berlin: Springer Nature, 54-79.
20. King, M. (2013). *Fisheries Biology, Assessment and Management.* New Jersey: John Wiley & Sons, pp400.
21. Luis Conceicao & Amos Tandler. (2018). *Success Factors for Fish Larval Production.* New Jersey: John Wiley & Sons, 53-85.
22. M. Valdimarsson & FAO. (2003). *Responsible Fisheries in the Marine Ecosystem.* London: Food & Agriculture Org, pp426.
23. Mahmut Caliskan, Halil Kavaklı, Gül Cevahir Öz, & Birgul Ozcan. (2015). *Molecular Approaches to Genetic Diversity.* New York: BoD – Books on Demand, 47-75.
24. Manfred Reinecke, Giacomo Zaccone, & B. G. Kapoor. (2020). *Fish Endocrinology (2 Vols.).* Florida: CRC Press 441-505.
25. N. Shimizu, T. Aoki, I. Hirono, & F. Takashima. (2013). *Aquatic Genomics: Steps Toward a Great Future.* Berlin: Springer Science & Business Media, 1-84.
26. Overturf, K. (2009). *Molecular Research in Aquaculture.* New Jersey: John Wiley & Sons, pp408.

27. Paul J. B. Hart & John D. Reynolds. (2008). *Handbook of Fish Biology and Fisheries: Fisheries.* New Jersey: John Wiley & Sons, 1-11.
28. Ray, S. (2018). *Biological Resources of Water.* New York: BoD – Books on Demand, 3-75.
29. Richard J. Beamish, & Brian J. Rothschild. (2009). *The Future of Fisheries Science in North America.* Berlin: Springer Science & Business Media, 1-77.
30. Robert DeSalle & Michael Yudell. (2020). *Welcome to the Genome: A User's Guide to the Genetic Past, Present, and Future.* New Jersey: John Wiley & Sons, pp280.
31. Zargar, R. K. (2020). *Rediscovery of Genetic and Genomic Resources for Future Food Security.* Berlin: Springer Nature, 1-66.

INDEX